Adhesives in Engineering Design

W. A. Lees
Technical Director, Permabond Adhesives Ltd
Woodside Road, Eastleigh, Hants SO5 4EX

The Design Council
London

Springer-Verlag
Berlin Heidelberg New York Tokyo

Adhesives in Engineering Design
W. A. Lees

First edition published 1984 by
The Design Council
28 Haymarket
London SW1Y 4SU

Typeset by Apex Computersetting, London

Printed and bound in the United Kingdom by
The Pitman Press, Bath

British Library CIP Data

Lees, W. A.
   Adhesives in engineering design.
   1. Adhesives
   I. Title
   620.1'99   TA455.A34

   ISBN 0 85072 150 4 The Design Council London
   ISBN 3 540 15024 2 Springer-Verlag Berlin Heidelberg New York Tokyo
   ISBN 0 387 15024 2 Springer-Verlag New York Heidelberg Berlin Tokyo

Contents

Acknowledgements

The author wishes to thank the management and staff of Permabond Adhesives Ltd for their unfailing support during the preparation of this book. The assistance given by colleagues in National Adhesives and Resins Ltd during the checking of documents proved to be invaluable and was greatly appreciated. Finally, the encouragement and help given so freely by Mr E. Ross, Engineer, during the preparation and checking of the manuscript must also be acknowledged. May I thank you all.

W. A. Lees
Eastleigh, August 1984

The ECV 3 is probably one of the most innovative cars ever built. A significant and unique feature is its aluminium monocoque body which relies entirely on a toughened epoxy-based adhesive for its structural integrity. (Photograph courtesy BL Technology Ltd; adhesive supplied by Permabond Adhesives Ltd.)

Introduction

The possible use of adhesives in a new design should always be considered because of the economic and technical benefits that they can confer.

Light, stiff and economic structures, free of the blemishes caused by conventional assembly methods, can be fabricated from a variety of materials in combinations which would otherwise be hard to achieve. Similarly, mechanisms may be built up using bonding techniques which are free of the costs and stresses implicit in press fitting.

Adhesives are not a panacea, but they do have a great deal to offer as is shown by the vital role they play in modern production engineering. Yet, despite this, they are not generally regarded with enthusiasm by engineers and designers. The reason for this is not hard to find. There are so many adhesives with such diverse properties that, in the absence of a unifying science which can explain not only why adhesives stick but why they behave as they do, a very strong incentive is required to guarantee perseverance. In addition, although the polymeric structures of adhesives are well understood, this knowledge is usually of little help to the engineer who is used to dealing in precise terms and may be readily put off by a subject which he tends to regard as being arcane and wooly.

Nevertheless, from the point of view of the designer, one of the most inviting avenues in his search for lighter and more economical structures and mechanisms leads inevitably towards the use of adhesives. The aircraft industry is one which has pursued this approach and has done so by relying on a relatively small group of materials and the incremental development of their associated technology. A similar determination has paid off for some forward-thinking companies in the motor industry, as is shown in the increasing use of adhesives in the construction of vehicles.

It is clear, then, that the resistance of traditional engineering practice can be and is being slowly broken down. Adhesives have been specifically developed for the mechanical engineer and the latest generation of 'toughened' adhesives have demonstrated quite clearly that they can make a major contribution to the work of both the structural and the mechanical engineer. But perhaps even more important than these developments is the growing understanding of adhesives as engineering materials in their own right. Their natural intransigence is yielding to a combination of mathematical modelling techniques, computer power and the development of sophisticated experimental techniques. In addition, the publication by the Ministry of Defence of a number of specifications covering anaerobic adhesives has eased the problems associated with the assembly of mechanical components to a considerable degree – a subject which is covered in some detail in Chapter 3. Elsewhere the background provided, coupled with shrewd and careful enquiry, will soon reveal what is possible.

The immediate object of this book is to provide the designer or production engineer with information which will allow him to make a rational decision regarding the possible use of adhesives. The tried-and-tested software package associated with this book further eases the process of decision making by guiding the user through a series of practical steps, each of which eliminates unsuitable adhesive types until only the appropriate alternatives remain. The absence of hard quantitative data in the text will no doubt come as a surprise to many, but their omission is deliberate. With few exceptions most of the data available are based on standard tests, but, as many of these allow considerable variation (typical being the choice of materials employed), it can be safely assumed that data emanating from different sources cannot be compared, let alone used for the purpose of a quantitative design.

1   Adhesives – their use and function

1.1 The use of adhesives

Although there is, as might be expected, a grey area between structural and mechanical applications of adhesives, it helps understanding to consider them separately. Generally, structural adhesives are used in some form of overlap joint, as distinct from the co-axial joints formed from turned parts, where anaerobic adhesives (see Chapter 3 and Section 5.1.2) are normally employed. However, structural adhesives may be used in co-axial configurations when maximum performance is required.

In addition to these applications, where severe loads are likely to occur, there are many situations in which an adhesive will be only lightly loaded – and where other modern materials may be employed.

It must be realised that, among the many roles of adhesives, their being used to secure mechanical components is a special one. It is often necessary to dismantle mechanical assemblies for maintenance and this requires the use of the unique anaerobic adhesives. These are strong enough to hold parts together during use, but not so strong that they cannot be dismantled with conventional tools and techniques.

Materials used in this way are known as adhesives despite the fact that, in the main, they are really very poor adhesives whose function is to 'jam' or mechanically lock parts together. For this they need very high, though adjustable, levels of compressive strength, and viscosities that enable them to fill the smallest interstitial space.

Generally, only the strongest materials in this group have adhesive characteristics. For the majority of applications 'jamming' is enough. Indeed, when they are used as gasketting media – an increasingly important role – adhesive strength can be a positive drawback.

Because these materials 'jam', cleanliness of the component parts is not of great importance, and even slightly oily parts may be successfully assembled. Naturally performance suffers but usually the reduction is quite acceptable.

Major adhesive family types are reviewed in detail in Section 5.1, while the selection procedure in Section 5.2 suggests the most appropriate adhesive family or sub-group for different applications.

1.2 How adhesives function

Atoms and molecules exert an attractive force on one another analogous to that of a magnet. This force, which is generated by the movement of electrons, is displayed by all substances – solids, liquids and gases – and in aggregate gives solids and liquids their everyday appearance. Because the force falls off rapidly with distance, separate particles will not adhere unless they can get close enough for these forces to come into play.

A single volume is easily produced from two separate portions of liquid, but not from two separate solids. Pour one portion of water into another and immediately a single liquid is obtained; but placing one block of steel upon another does not create a single block. This is because the individual surfaces are too inflexible to approach closely enough for the atomic forces to pull them together. Very highly polished steel blocks do tend to stick together and display some degree of adhesion because between very smooth surfaces there will be close interaction between a few – very few – small, isolated areas.

Some materials, notably some rubbers and gums, behave both like liquids and like solids, the distinction between the two forms being a function of time. Because the materials are soft and flow very slowly they behave as solids when first placed one on top of the other but, given time, the two surfaces will grow into a single solid mass. This is the action of an adhesive.

Being liquid, adhesives flow over the surface of a solid and, because of their intimate contact, interact with its molecular forces. Then, as a result of the adhesives' own curing processes, they become strong solids which, retaining intimate contact with the surfaces, hold them together. The usual need for surface cleanliness is now readily apparent: unless the adhesive can dissolve residual oil and grease, close contact with the underlying surface will not occur and the joint will be poor. A loose deposit of surface debris has the same effect: the joint will be weakened because inter-molecular forces cannot come properly into play. The general performance of some modern adhesives has been enhanced by their ability to dissolve surface contamination and absorb solid debris.

As no adhesive is as strong as metal, the adhesive interlayer will always tend to be the weakest link in a bonded structure. Care is therefore needed to ensure that service stresses are well within its capabilities. This is normally easy to achieve by providing relatively large areas in a joint.

A more formal review of the nature of the various forces which control adhesion is to be found in Appendix 1.

It is important to understand why adhesives fail as well as why they succeed, and to appreciate the factors that lead to their degradation. There are many causes, but a major one is the brittleness of the traditional high-performance adhesive. To achieve very high shear strengths – and therefore the capacity to sustain high loads – it has been necessary to produce very hard and brittle bonds. Regrettably, such bonds are particularly prone to damage when the structure is shock loaded and the momentary distortion of the substrate generates powerful peel and cleavage forces that they may not be able to resist (see Figure 1.1).

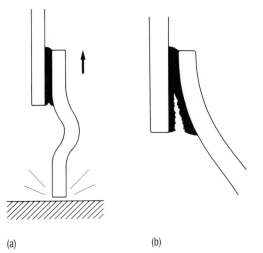

Figure 1.1   Peel and cleavage forces induced by shock loads. a A bonded object striking a rigid surface experiences powerful and distorting shock waves. b Depending on a variety of factors this distortion may cleave the adhesive apart, particularly if the adhesive is brittle.

(a)   (b)

This tendency for an overloaded joint to fail catastrophically when distorted has been substantially supressed by the introduction of 'toughened' adhesives, which inhibit crack propagation by having a dispersed 'rubbery' phase within the harder load-bearing adhesive matrix (see Figure 1.2 and Plate 1). When the joint is overloaded, the force that would induce and propagate a crack is dissipated in the flexible rubbery phase. A similar strategy is used when stopping a crack in a sheet of over-stressed metal or plastic by drilling out its tip. Instead of fracture energy concentrating at the crack tip, it disperses around the periphery of the drilled hole.

Although development of 'toughened' adhesives has been limited to the various acrylic (both anaerobic and non-anaerobic) and epoxy types, overall performance has been improved spectacularly. Both environmental resistance and peel and impact performance have been raised substantially without sacrificing shear strength.

Figure 1.2   In an overloaded toughened adhesive, crack propagation is stopped by the dispersed rubbery phase.

2 Design

2.1 Introduction

A typical joint cross-section is shown in Figure 2.1, where a clear distinction is made between the bulk of the adherends and their surfaces. Chapter 4 discusses the importance and nature of a component's surface and methods of improving its structural strength and increasing its suitability for bonding. However, it must be borne in mind that sophisticated surface preparation is often necessary only for major load-bearing structures – often using some form of lap joints – which are exposed to severe environments.

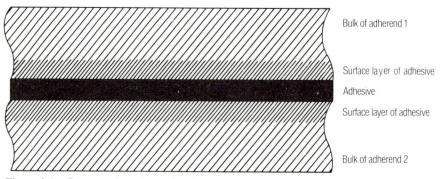

Bulk of adherend 1

Surface layer of adhesive

Adhesive

Surface layer of adhesive

Bulk of adherend 2

Figure 2.1   Cross-section of a typical joint. Note that the thickness, density and form of the surface layer of each adherend varies with its nature and history. The picture is complicated if the adhesive dissolves, permiates or reacts with the surface.

Adhesives for the materials and surfaces to be bonded are best selected by the procedures given in Section 5.2.

Problems the designer must still consider include:

Loads to be borne by and distributed within the various layers of the joint shown in Figure 2.1; and

Practical aspects of assembly.

In both cases, a distinction must be made between the many variants of lap joints on the one hand and co-axial assemblies in general on the other. Threaded components must be singled out, for, when bonded, they do not behave as slip-fitted unthreaded parts. They are therefore dealt with separately in Sections 2.3.1 and 2.3.2.

2.2 Basic considerations

2.2.1 Introduction

The various types of joint assembly fall into three distinct categories:

Mechanical components – usually turned and fitted.

Loaded structures – usually bonded sheet metal, though GRP is increasingly used.

Nominally loaded small parts and trim.

The last of these categories poses few major problems and the following rules apply chiefly to the design of loaded assemblies.

2.2.2 Joint area

If significant loads are to be borne then adhesive manufacturers' performance data will give a rough guide to the bond area required, which will have to be refined considerably in the light of the following two sections.

2.2.3 Safety factors

In choosing an adhesive and calculating the bond area required, designers should take into account not only the base load but also fatigue, creep and environmental stress, not forgetting loads likely to arise from differential thermal expansion.

To a certain degree, bond area can be increased to accommodate workmanship

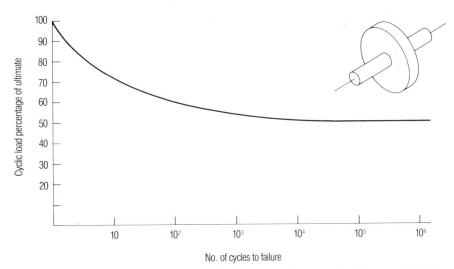

Figure 2.2   Fatigue performance of a typical toughened anaerobic adhesive. This S/N curve (load v cumulative number of loading cycles to induce joint failure) for a toughened anaerobic adhesive shows that it will not fail through fatigue if the load applied is less than 50 per cent of the ultimate strength in shear of the adhesive – in this specific joint configuration.

problems and unavoidable variations in component dimensions, surface finish and condition. But these factors can also be controlled by employing the right adhesive, good application equipment or techniques and a good quality-assurance system.

Materials sold as load-bearing structural adhesives may have quoted fatigue performances as high as 50 per cent of their Ultimate Tensile Strengths (UTS) – see Figure 2.2 – but these relate to idealised conditions and, as a rule, adhesives should never be loaded beyond 10 per cent of their UTS.

Aircraft joints are rarely loaded beyond 5 per cent of UTS. At these stress levels, high-quality adhesives cope well with most environments and will not creep.

Finally, fatigue may be substantially reduced by ensuring that the loads are not reversed (see p.22 Fatigue).

2.2.4 Stress distribution

Tensile loads normal to the plane of a joint can be highly destructive and should be avoided if possible. Indeed, a joint ought to be designed to carry a compressive element within its load pattern (see Figure 2.3, Section 2.3.3 and, in particular, Section 2.3.3.7).

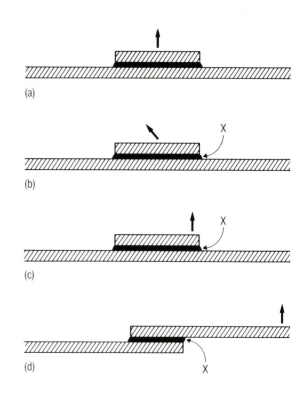

Figure 2.3   Tensile forces normal to the plane of the joint can become highly destructive if the design deviates from the ideal situation (shown in a). b An inclined tensile load induces cleavage at X which is enhanced if the adherend is long and can act as a lever (see d). c Cleavage also occurs if the load is off centre and is accentuated (d) if the tensile butt joint format develops into a lap joint.

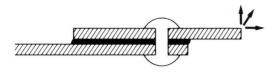

Figure 2.4   Peel or cleavage forces may be suppressed by the use of local mechanical restraints, such as rivets, bolts or spot welds.

Even the best of the latest generation of toughened adhesives can carry loads roughly 100 times greater in standard shear tests than they can in peel. While such tests are not directly comparable, they show that even small peeling loads are particularly destructive.

In contrast, compression loads are readily borne. If these are compared, then the relative ability of a joint to withstand compression, shear and peel loading is of the order of 1000:100:1.

Clearly, a joint should be designed to avoid peel. However, if peel and cleavage forces must be borne, then some means of distributing the load must be found, for example by restraining movement at the end of a flat, flexible joint with a rivet or spot-weld (Figure 2.4). If such remedies are not practicable, then complete re-design may be required.

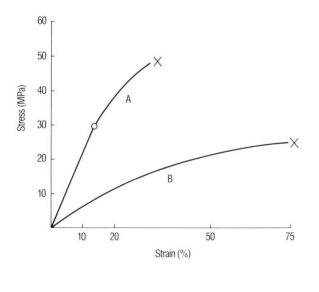

Figure 2.5   Stress/strain characteristics of two toughened adhesives. A is a toughened, single-part epoxy with a clearly delineated elastic limit (0). B is a highly compliant, toughened acrylic where the limit of the elastic zone is not clearly defined. Both adhesives fail at X. As the temperature rises, B flattens further and A takes on the B form. As the temperature falls, A becomes steeper and B takes on the A form. By choosing the correct formulation the modulus appropriate to the application may be obtained (see Figure 2.6).

Another common error is the belief that the load on a joint is uniformly distributed and represented by the formula: Load per unit area = Load/Area

Even with the most compliant adhesives, this ideal state never occurs. Furthermore, the stress distribution within a joint alters as the moduli of the adhesive and the materials being bonded vary with temperature changes. Figure 2.5 illustrates this for two adhesives, one of which has a high level of compliance while the other has not. How temperature affects stress distribution within a joint as is shown in Figure 2.6.

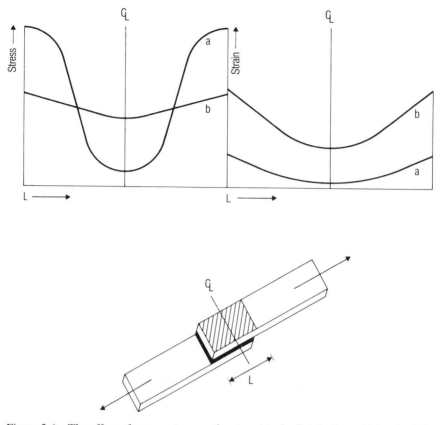

Figure 2.6   The effect of temperature on the stress/strain distribution within a lap joint. As the temperature rises and the adhesive softens and the initial load pattern changes. Greater distortion at the edge of the joint distributes load by transferring it to the central areas. As more of the inner area becomes stressed with continued temperature rise, the adhesive ultimately begins to creep. The rate increases as conditions become more extreme. Failure is imminent in curve b. (See also Figure 2.7.)

Computer programs are now becoming available for designers to examine the interplay between the moduli of materials, joint geometry and dimensions. Figure 2.6 was developed from just such an evaluation.

Without such specialised programs, it is safer to assume that, for sheet metal approximately 1 mm thick (16-18 gauge), all the load will be carried by the first 6 mm or so of each end of the overlapped area – in other words the minimum overlap for any form of load-bearing structure should be at least 12 mm. Once overlap exceeds 12 mm, its central area ceases to be significantly stressed. When such an overlap is used with a high-performance structural adhesive, the joint's performance and ultimate failure will be significantly affected by the adherend's modulus. It is often difficult, or impossible, to say which is responsible for the joint's ultimate rupture as the load rises to an intolerable extreme.

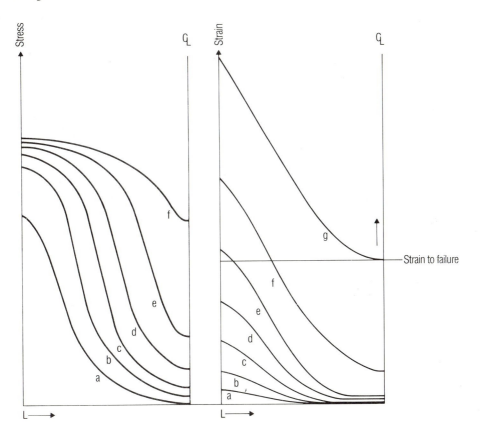

Figure 2.7 Stress/strain distribution in a steel/steel (1 mm) lap joint subject to a variety of stress level. As the load on a lap joint bonded with a high modulus, toughened adhesive is increased uniformly (a-f), there is a rapid and disproportionate redistibution of strain and stress. At load e, when the failure strain is reached, distortion increases very rapidly and total collapse is imminent. It is here that the differences between toughened and untoughened adhesives are most marked. Typically, untoughened adhesives fail catastrophically at e; but, despite the damage at the edge of the joint implied by the transition of the adhesive in this area through γ_F, toughened adhesives will not do so – providing L is at least 19 mm (see Figures 2.8 and 2.9).

Figure 2.7 shows the stress distribution in a typical lap joint subjected to a variety of loads. Figure 2.8 compares the stress changes generated by a common load in a variety of lap joints and Figure 2.9 highlights the effect of this on the strength of the joint.

Thus the strength of a joint cannot be improved simply by increasing the overlap length. This can only be done by changing parameters such as the width of the joint, the thickness or nature of the adherend or substituting a more appropriate adhesive.

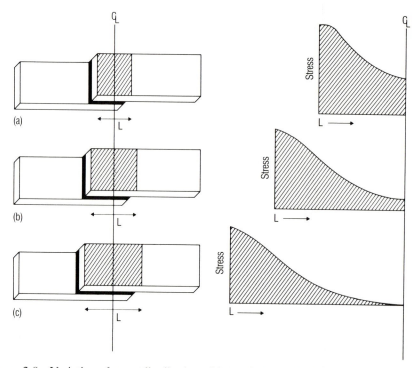

Figure 2.8   Variation of stress distribution with overlap length. As the overlap length of the lap joint is increased, the maximum stress observed (at the edge of a joint) for a given load falls: hence c is a stable joint whereas, at the same load, a is on the point of failure. However, a further increase in the joint overlap (Figure 2.9) does not bring any further reduction once the central area of the joint is no longer loaded – though that the latter should be so is essential to avoid creep failure. The joint c can only accept a further increase in load (Figure 2.7) at the expense of the unloaded area a its centre, which puts the joint a risk because of the increased probability of creep and possibly fatigue failure. Higher loads may only be safely carried if the width of the joint is increased.

Brittle adhesives behave very differently from toughened adhesives when the joints they form are oveloaded. Because they do not crack, overstrained tough adhesives fail progressively – the stress pattern in the joint moving through the stages represented by c, b and a. By contrast, the brittle adhesives fail totally and catastrophically.

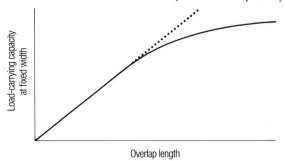

Figure 2.9a

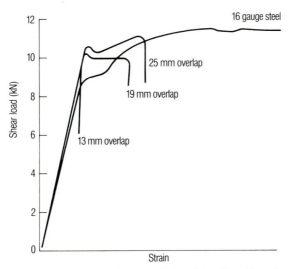

Figure 2.9   In a (*opposite*) the proportionate relationship of load-bearing capacity with overlap length falls away (at about 10-15 mm in Figures 2.7 and 2.8). The actual performance of a rubber-toughened, single-part epoxy is given for a series of lap joints in b . The strength of the various joints, based on 16 gauge (approx 1 mm) steel is compared with the stress/strain performance of the steel itself.

The simple rules of thumb below that may be made for the thinner sheet metals may not be valid for thick sections.

The design of many relatively simple joints that will not be subjected to extremes of load or environment may be based upon the following empirical rules:

> Minimum overlap: 12 mm (5-6 mm for nominal loads).
>
> Maximum overlap: 25 mm (this allows for accidental damage either within or alongside the joint area).
>
> Base the load-bearing area on the recommended overlap length and the appropriate width or, in the case of co-axial components, diameter.
>
> Design the joint to minimise peel or cleavage forces, and consider the possibility that these may arise as a result of accidental damage.
>
> Specify the maximum rigidity or thickness of component compatible with other cost/weight considerations.
>
> Maximise fatigue resistance by ensuring that shear loads do not oscillate between compression and tension.

No account has been taken so far of the thickness of the adhesive within the joint line, but conventional wisdom , based on standard test procedures, dictates a bond line that is as thin as possible. While, to a degree, this is equally true for both the older types of conventional high-strength – though brittle – adhesives and the new 'toughened', non-brittle adhesives, the peel strength of the latter increases with bond thickness, as shown in Figure 2.10. And, it should be borne in mind that real joints are not usually stressed like standard test joints.

This is of great practical importance since it is peel and cleavage forces induced by accidental impact that usually destroy joints (see also Section 2.3.3).

Figure 2.10   Effect of bond line thickness on the peel strength of a toughened epoxy-based adhesive. Although not directly proportional, the effect is substantial and of considerable practical importance because bond line thickness is often difficult to control. In most cases the overall result of an increase in bond line thickness is beneficial, even though the shear strength may fall; most joints fail because they cannot meet peel and cleavage overloads. Generally shear overloads are rare.

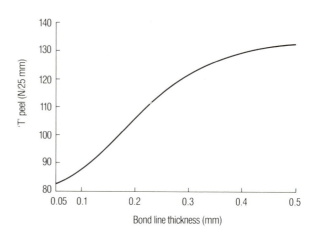

2.3 Practical aspects

Apart from the physical problem of actually assembling components sufficiently well to make an effective joint, a number of practical considerations need to be taken into account. For example, where two components are to be assembled as a lap joint, the deliberate shaping of one part – perhaps with a slight indentation – may allow the more accurate placement of the second. Similarly, joint strength will be enhanced by ensuring that the load is borne in compressive shear rather than trying to design for pure shear loading – which is often virtually impossible.

Elastic movement of the components will always result in distortion and this, if not considered, may result in peel and cleavage.

Finally, care must be taken, either in the choice of adhesive or in the assembly technique, to ensure that the adhesive is not wiped from the surface as the parts are placed together. This is likely where, for example, one part is slid into position over another; the sliding action tends to wipe the adhesive from the joint area, forming pockets and reducing overall performance. In this context, co-axial assemblies must of course be slid together because of the very nature of the joint. This can cause severe problems for some adhesives but presents no difficulties for the anaerobic varieties, which are formulated specifically to cope with this type of work.

Although the selection procedure in Section 5.2 will eliminate adhesive families and sub-groups that are unlikely to meet the basic requirements of any particular design, the designer should take this process further, preferably with the assistance of a manufacturer of the selected type of adhesive, by identifying or specifying an individual formulation that will most readily meet production requirements. The most suitable adhesives will:

Store well;

Not require mixing;

Dispense readily through simple equipment;

Cope well with inevitable variations in production conditions;

Resist, in either the uncured or cured state, the damaging effects of subsequent assembly processes; and

Be safe to handle.

2.3.1 Co-axial assemblies (excluding threaded parts)

The many joint configurations based on the smooth, unthreaded collar and pin assembly can be grouped as follows:

2.3.1.1 Permanently assembled components

Typical components are:

Cams	Knobs	Rotors
Cylinder liners	Pins	Shafts
Fans	Pipes	Sleeves
Gears	Pulleys	Splines
Impellors	Rings	Tubes

Bonding these components can bring numerous benefits that are often quite unattainable with conventional assembly techniques. These are discussed below and also in Chapter 3.

Until computer-aided design programs are freely availabe, together with the necessary data, it would be unwise to rely on a power transmission design that loads the adhesive in shear rather than compression without extensive practical trials (see pp.16 & 22 Supplementary role of adhesives in power transmission and p.33 Splines, keys and set screws).

With this restraint in mind, the most likely order of preference – taking general robustness into account – would be toughened adhesives of the following classes: Single-part epoxies, Acrylics, Anaerobics, and Two-part epoxies

Each group has specific design limitations and freedoms.

Toughened single-part heat-cured epoxies
The general purpose versions of these adhesives invariably employ a solid catalyst and so it is never safe to assume that they may be used in gaps of less than 0.05 mm (or 0.1 mm on the diameter). Furthermore, the means of applying them must be carefully evaluated. Methods for consideration include injection, heat-induced capillary flow from a shallow well or groove, or design and placement techniques based on Figures 2.11 and 2.12, which have been specifically devised to fill the joint completely and avoid entrapped air. Care must also be taken to ensure that the design will not allow the adhesive to flow out of the joint during the heating cycle, perhaps by incorporating a snug-fitting shoulder to close the lower joint face.

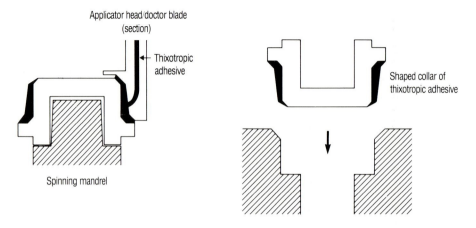

Figure 2.11   Correct placement of a component pre-coated with dispenser-applied adhesive. a Thixotropic adhesive is placed on the plug by means of a spinning mandrel coupled with a shaped application head and doctor blade. b Air will be displaced on plug insertion – given the illustrative dimensions of Figure 2.12. This will ensure that the adhesive will be uniformly distributed and that there are no leakage paths.

Some versions of this adhesive type are very viscous throughout the temperature range of their curing cycle and will neither creep nor flow during the hardening process. Thus, provided that they can be successfully placed in the narrowest parts of the joint, these materials will generally cope with any gap.

In very demanding situations where maximum performance is required, additional strength can be gained by reducing the adhesive's internal stress as a result of minimising differential expansion effects through curing at the most appropriate temperature. This will generally be the manufacturer's lowest recommended cure temperature, but because of other factors specific advice should be sought.

Toughened acrylic adhesives

These compliant adhesives offer the great advantage of room- temperature curing and consequent freedom from residual thermal stresses. However, from the designer's point of view, they suffer from two potential disadvantages which, according to the actual adhesive, may be experienced either alone or together.

Many of these adhesives have a rather stringy nature and all require the use of a cure initiator, which may be pre-mixed prior to application or applied to at least one of the adherends prior to assembly.

Two techniques to overcome the problems stemming from the combination of the adhesive's stringy character and the initiator's slipperiness are:

Tapering the components to give a cone and socket assembly, which is very successful for pre-applied initiators. Tapering can also improve load distribution, though an engineering model and computer assistance will almost certainly be needed to exploit the situation fully.

Injecting pre-mixed formulations.

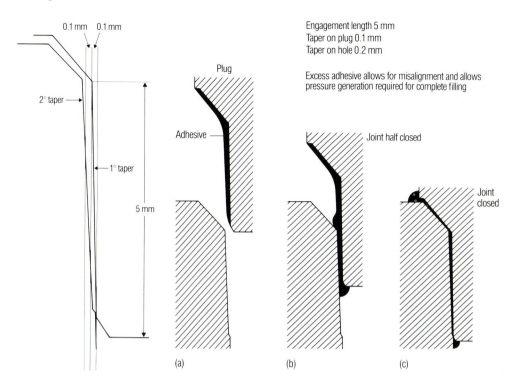

Figure 2.12   a Shape and dimensions of plug and socket for complete sealing. b Inserting plug into socket: when half inserted the pressure in the taper forces the adhesive both forward and backward, ensuring that the gap is completely filled.

Because a number of variants of the basic adhesive have been developed (though the range is still relatively limited compared with the older systems) materials are now available to cope with almost any clearance likely to be met – even the extreme case of filling the minute gaps of an interference fit.

Toughened anaerobic adhesives

Adhesives of this type are easy to use – they require neither heat nor a second component to cure them – and always merit consideration. They are, however, expensive and their inability to cope with really large gaps often prevents their use on large structures. There are no viscosity-related problems with products of this type. The limiting feature is the surface polymerisation process which restricts adhesives of this type to gaps of approximately 0.25 mm.

Anaerobic adhesives are not generally stringy in nature and contain no solid catalysts, so they are suited for application on many relatively small parts – particularly co-axial assemblies where they can be placed with ease. Their characteristics ensure that they readily fill the minute gaps always found in interference-fitted joints. Thus, they can be used both to seal and supplement the overall performance of such joints.

Toughened two-part epoxies

Warming these cold-curing, resinous liquids and pastes reduces their viscosities (and increases their cure rates) although this is not normally convenient before the adhesive is placed. A practical lower gap limitation for unfilled versions should be taken as approximately 0.025 mm (0.001 in.).

Supplementary role of adhesives in power transmission

Where power delivery would impose an excessive shear load on an adhesive, joints should be designed to load the adhesive in compression, as in spline or key assemblies. Despite the extra work required in their fabrication, using an adhesive to retain such components is still a better choice than the traditional interference fit.

An anaerobic adhesive is the best choice for these applications and only under extreme environmental conditions is it necessary to resort to the toughened variants or – in the final resort – the toughened, single-part epoxies (see p.33 Splines, keys and set screws and p.71 Co-axial – splined).

On joints carrying some torque load but not delivering power, the normal anaerobics cope well and their ease of use and versatility makes them the natural choice. Generally, only extreme environments, large gaps or other considerations dictate selection of the toughened variants and the single-part epoxies. These impose other design restraints (see Toughened single-part heat-cured epoxies and Toughened acrylic adhesives, above).

2.3.1.2 Dismountable components

Very often, bearings, bushes, splines, keys and other co-axial assemblies must be dismantled for replacement or maintenance and here the only practical material for use in the original assembly is an anaerobic adhesive (see Chapter 3 – in particular Section 3.6).

2.3.1.3 Specific applications

The general case – collar and pin

The general case is represented by the collar and pin shown in Figure 2.13. To ensure ease of assembly and reliable service the designer should consider the following points:

    Gap
    Alignment
    Strength
    Fatigue
    Heat
    Joint area
    Assembly
    Surface cleanliness
    Plastics and soft alloys

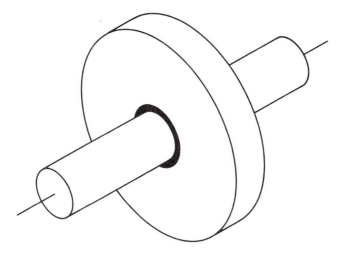

Figure 2.13   The collar and pin joint.

Gap
The dimensions discussed in Table 2.1 relate to the gap-filling capacity of the various adhesive families, not to the engineering function of the assembly.

For precision engineering applications, clearly the most likely choice is anaerobic adhesives.

Table 2.1   Gap filling capacities of adhesive types likely to be used in co-axial assembly

Adhesive type	Gap-filling capacity (based on diameter)
Anaerobic – all types	Interference to 0.25 mm (0.01 in.) according to individual formulation
Cyanoacrylate – all types	From 0.0025 mm (0.0001 in.) to 0.05 mm (0.002 in.). Larger gaps not recommended unless 'accelerator' and higher viscosity varieties used
Epoxy – single-part (heat-cured)	Minimum recommended 0.05 mm (0.002 in.). No practical upper limit
two-part	Minimum 0.025 mm (0.001 in.). No practical upper limit
Toughened acrylic – unmixed	According to formulation from a lower limit of approximately 0.0025 mm (0.0001 in.) to 0.25 mm (0.01 in.)
pre-mixed	Lower limit approximately 0.0025 mm (0.0001 in.) according to type with no practical upper limit

Alignment
Jigs are essential for precise axial alignment of components being bonded. In practice, axial displacement errors in bonding are much more acceptable than the angular error produced by a combination of off-set bores and insertions. The extent of both of these effects is examined in Figures 2.14 and 2.15.

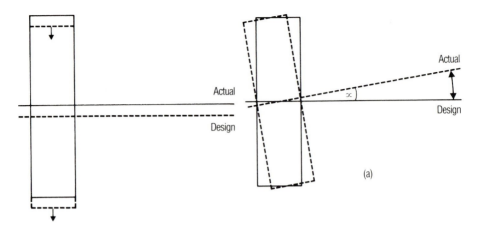

Figure 2.14   (*Left*) Displacement error caused by slip-fit and bonded assembly. When components are slip-fitted, the normal maximum axial error is dictated by the allowed tolerance and the intended separation – usually 0.025-0.05 mm. However, press-fitted parts are subject to angular error, which produces the exaggerated effects which can be seen in Figure 2.15 (*right*).

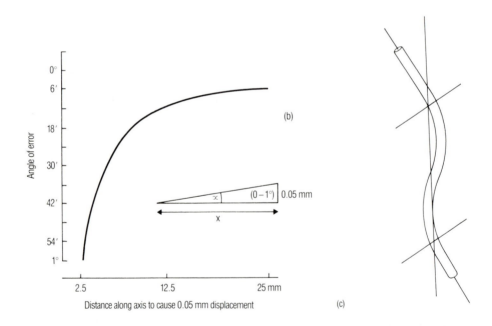

Figure 2.15   If a press-fitted component is subject to angular misalignment (a) then the maximum error normally seen in a slip-fitted assembly is readily exceeded (*top right*) (b). Where a second bearing is involved, shaft distortion can occur (c). (See also Figure 2.20.)

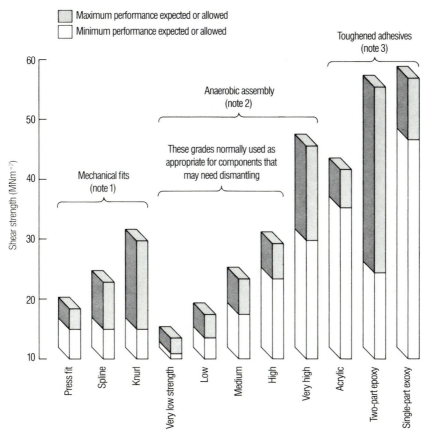

Figure 2.16   A comparison of the degree of retention displayed by mechanical techniques, anaerobic adhesives – which essentially function by jamming components – and several true adhesives.

Note 1: The lowest level of retention is a function of use, not technique. The higher levels may well involve considerable fabrication costs, greater accuracy and yet still leave parts stressed and prone to mechanical problems.

Note 2: The maximum and minimum performance levels are dictated by the requirements of the MoD performance specifications DTD 5629-33 inclusive – except in the case of 5633 where no upper limit is set. In the diagram the usual performance figure is given (see Figure 2.13).

Note 3: Normally adhesives (non-anaerobic) are not assessed on collar and pin joints – but for comparison with the mechanical and anaerobic techniques the performance of several typical adhesives has been assessed using this method. Acrylic: This type of toughened adhesive does not show up well on joints of this size because it is too compliant to be effective on small joint areas. On a large lap joint, a 50 per cent increase could be anticipated. Two-part epoxy: The many formulations available could fall anywhere in the band indicated according to their individual characteristics. Single-part epoxy: These toughened adhesives offer the maximum performance possible and are usually easier to use than two-part types.

Strength

Figure 2.16 compares the relative co-axial shear strength (measured on the Collar and Pin test specimens described in MoD DTD 5628, Method H) of a number of adhesives with a resistance to disassembly similar to related press-fitted collar and pin components. The strength of the bonded joints is considerably in excess of the resistance to movement displayed by friction- fitted assemblies.

Anaerobic adhesives offer an additional advantage in being available in a specific range of strengths based on the requirements of the MoD's DTD 5629-33 inclusive. So bonded assemblies may be made suitable for disassembly using conventional workshop equipment, or made permanent.

The adhesive should be selected with the end user and maintenance in mind. A very strong adhesive naturally confers a greater margin of safety, but this is often unnecessary, and the indiscriminate use of very strong grades may cause problems if dismantling is needed.

The stress analysis of bonded joints remains in its infancy and, until it is better developed, simple shear strength values such as those given in Figure 2.16 (and typically provided by most manufacturers) should not be used alone to assess the ability of an adhesive to transmit power (see pp.16 & 22 Supplementary role of adhesives in power transmission).

Other mechanical characteristics, such as the moduli of the adherends, the geometry of the joint, fatigue and environmental exposure, must also be considered.

Although the data of Figure 2.16 cannot be used to predict adequately an adhesive's performance over a prolonged period, it can be said, as a very broad rule, that toughened adhesives will perform better than untoughened ones. Again without appropriate design models, the position may reasonably be summarised as follows.

If a conventional, mechanically assembled joint would normally rely on an interference fit, then an adhesive of equivalent strength (see Figure 2.16) would be a sensible substitute using slip-fitted components. However, if the design calls for a pin, key or spline to transmit very high loads or power, then an adhesive, used in conjunction with them, will extend their useful life and offer economies through relaxed tolerances.

The actual shear strength displayed by a bonded collar and pin assembly depends on diametric gap and surface roughness. Figures 2.17 and 2.18 indicate how these affect the performance of a typical joint bonded with anaerobic adhesive. However, while economies achieved through relaxed tolerances and surface finish must not be overlooked, these variations are usually of little importance in terms of the practical performance of the assembled components, because in most co-axially assembled components the load is essentially compressive. Substantial axial loads are rare, and the rotational shear loads imposed by the torque of an electric motor rotor – typically assembled on to its shaft with anaerobic adhesives – are quite low. Again, power should only be delivered by shear loading after very careful design or prolonged testing (see

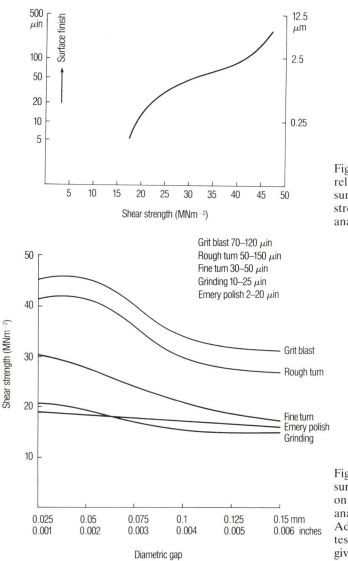

Figure 2.17 Idealised relationship between surface finish and shear strength for a high-strength anaerobic adhesive.

Grit blast 70–120 μin
Rough turn 50–150 μin
Fine turn 30–50 μin
Grinding 10–25 μin
Emery polish 2–20 μin

Figure 2.18 Effect of surface finish and gap size on the shear strength of an anaerobic adhesive. Adhesive DTD 5633/G2 tested to DTD 5628 using given finish and gap.

pp.16 & 22 Supplementary role of adhesives in power transmission).

Anaerobic adhesives of appropriate viscosity are often used to augment and standardise the performance of shrink and interference-fitted components, where they can reduce hoop stress in the outer component and reduce its tendency to crack or display fretting corrosion. These benefits stem from the ability of anaerobics to fill the voids that are present in even the best-prepared surfaces. This spreads loads evenly over the whole of the faying surfaces instead of concentrating them on the engaged peaks, which represent only a small proportion of the theoretical contact area.

Fatigue

Fatigue is only likely to occur if power is being delivered in shear through the joint line, and it may be avoided by keeping the continuous running load to around 10 per cent of the static shear strength of the adhesive, measured on the joint concerned. This may equate with the UTS value (see Section 2.2.3 and Figure 2.2).

However, if the quoted UTS value comes from a different type of joint, it may well be a poor guide to anticipated performance. Hence an arbitary UTS figure should never be used to gauge fatigue performance – a point to be borne in mind in discussions with adhesive suppliers. Although shear loads may be nominal or low, a cumulatively significant number of higher loads may be imposed during the life of the assembly. For example, although a gear pump does not normally transmit very high torque loads, the inertial starting and stopping loads in some designs may ultimately cause fatigue failure.

Heat

The load-bearing capacity of an adhesive reduces rapidly as its T_g (glass transion temperature) is exceeded, so only a few individual formulations – even the toughened ones – can perform well when bond line temperature exceeds 120°C.

With co-axial joints, an increase in temperature does not necessarily cause the anticipated reduction in performance because the differential expansion of the adhesive and the assembled components may generate very high compression forces on the adhesive which in turn result in an increase in observed shear strength. It is not generally realised that typical adhesives have thermal expansion co-efficients between seven and twenty times those of common engineering metals, as shown in Table 2.2.

These data show how the simplified situation described above can become complicated if the bond line is very thin, or if the adherends are substantial or are made of different metals. A brief calculation readily shows that a large, light alloy

Table 2.2   Co-efficient of expansion of some common engineering materials

Substance	Co-efficient $\times 10^{-6}$
Aluminium	29.0
Brass	19.0
Bronze	18.0
Copper	16.7
Gunmetal, Admiralty	18.0
Iron, cast	11.0
Iron, wrought	12.0
Magnesium	25.0
Steel, mild	11.0
Steel, stainless	10.4
Titanium	9.0
Zinc	30.0
Typical anaerobic adhesive	210.0

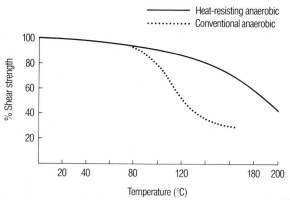

Figure 2.19   Effect of temperature on the shear strength of conventional and highly cross-linked anaerobic adhesives. The poorer performance of the conventional version as measured on a standard collar and pin assembly (MoD specification DTD 5628) is quite marked.

housing containing a steel inner component will readily overload the most accomodating adhesive.

Nevertheless, experience shows that, for the vast majority of like-to-like, co-axial assemblies with diametric clearances around 0.05 mm, temperatures between -55°C and +80°C will be readily accommodated. Adhesives with a higher T_g will cope up to 120°C without major strength loss, and those designed for elevated temperatures perform, on collar and pin assemblies, up to 200°C. This is seen clearly in Figure 2.19, which shows the performance of a highly cross--linked anaerobic adhesive specifically formulated for maximum performance on collar and pin assemblies at elevated temperatures. It cannot be emphasised too strongly that the data given relate only to collar and pin assemblies and must not be applied to lap joints.

The introduction of toughened adhesives has brought major improvements in performance, since they are not prone to brittle failure. Very high shear strengths are now attainable on both collar and pin and lap joint assemblies over an extremely wide temperature range.

However, it should be remembered that the vast majority of adhesives degrade rapidly as the temperature approaches and passes 200°C. Consequently, whenever a service temperature of 120°C or more is likely, the expected service life of the adhesive should be discussed thoroughly with the manufacturer.

Some adhesives, typically the polyimides, withstand long periods at temperatures well above 200°C. However, the use of such materials may cause practical difficulties as, for example, with a film-based adhesive in a co-axial assembly.

Joint area

Just as in the simple lap joint (see Section 2.2.4 and Figures 2.8 and 2.9), joint strength, in terms of axial shear strength, does not increase progressively with joint area because the load is carried on the leading and trailing edges of the joint. Typically, only a very small – probably insignificant – proportion of the load is carried by any part of the joint more than 5-6 mm from the edge normal to the imposed load, unless very ductile anaerobics are used.

As bond area is the product of width (diameter) and length, the strength of the joint will only increase significantly with width once the joint length exceeds some 10 mm. Once again, estimates of joint strength can only be very approximate until appropriate mathematical models and computer software become freely available.

Fortunately, in most applications, turned and fitted parts are usually compression-loaded, and any shear loads imposed – either by design or accident – are readily accommodated.

Assembly

The interaction of component design and the ease of placing an adhesive has already been mentioned, but the assembly of large, interference-fitted parts poses its own special problems. Generally, only anaerobic adhesives are suitable here and the adhesive's viscosity/thixotropy characteristics need to be tailored to suit the parts involved. Reaction rate, too, will almost certainly have to be reduced so that the heat generated during fitting does not cause premature polymerisation which would hinder, and possibly even prevent, assembly.

Shrink-fitting by cooling one component appears an attractive alternative as it avoids frictional heating. Unfortunately, frozen condensate usually forms on the cooled component and leads to unpredictable results.

The alternative – heating the outer component – is, however, completely viable and has been used to assemble huge components, as illustrated in Plate 3. Here, a gear shaft coated with anaerobic adhesive is being slipped into a collar pre-heated to approximately 100°C. The use of a line-to-line fit coupled with the adhesive has eliminated excessive hoop stress to give a strong compression joint; the adhesive acting as a space filler – so distributing load – and, to a lesser extent, in the true role of an adhesive to supplement the assembly's frictional strength.

Component heating or the use of slow-curing adhesives will solve most of the assembly problems of very large interference fitted parts. Another satisfactory approach is to use tapered components.

Surface cleanliness

Provided that co-axial joints will be loaded in compression – as they very often are – then cleanliness of the surfaces to be bonded is not too important.

While oil contamination will reduce the strength of a co-axial joint, this is normally of little importance unless power is to be transmitted through shear loading of the adhesive (see p.16 Supplementary role of adhesives in power transmission). Where power is delivered to keys and splines, cleanliness of the load-bearing surfaces is again relatively unimportant because power transmission relies on compression loading of the face or faces of the key or spline. Where maximum adhesive performance is required from any of the types suitable for co-axial assemblies, simple solvent washing to remove gross contamination may not be sufficient. Appropriate techniques are described in the general literature, although those for many of the common materials are summarised in Table 4.3 (Section 4.3).

Plastics and soft alloys

If one or more of the parts in a bonded assembly is made of a plastics material or soft alloy, then the overall performance will be lower than if the parts had been made of steel. It is important to remember that the shear strength of co-axial joints shown during the displacement of the parts is actually a measure of the bonding and mechanical jamming action of the hardened adhesive. The emphasis on either bonding or jamming depends upon the type of adhesive.

Thus, with plastics parts, two factors reduce the performance of the bonded joint. First, plastics are more flexible and reduce the jamming action. Second, bonding to many plastics is difficult and sometimes impossible. To some degree, this reduction in performance can be off-set by using high-strength, though ductile, adhesives and, if possible, by enhancing the mechanical contribution by increasing the roughness of the faying surfaces. The surface of the plastics may also be chemically treated to maximise adhesion. Again, see Section 4.3, Table 10.

Bearings and bushes

Anaerobic adhesives are often used for the quick, effective and economical fitting of bearings and bushes, obviating any risk of mechanical damage by press-fitting.

One or two individual grades of adhesive (see Section 3.6) cope adequately with most designs. Extremes of size and clearance are met by the many other grades, providing a wide range of viscosities and strengths.

Because anaerobic adhesives can make such a major design contribution in this area, it is worth examining the problems caused by conventional assembly techniques that are avoided by bonding.

Retention

To generate sufficient hoop stress to retain components by friction, the part dimensions for interference fitting must be very accurate to avoid any combination of the following problems:

   Intermittent movement causing galling and corrosion, possibly followed by either seizure or unwanted movement and inevitable premature failure.

   Excessive closure of bearing raceway clearances, leading to noisy running and premature failure.

   Immediate cracking – or rapid fatigue failure – of over-stressed rings and raceways.

   Scraping of inaccurately machined parts.

   Inaccuracy caused by fitting new parts on to worn components.

   Difficulties of retaining bearings in plastics or light alloy housings, especially at higher temperatures.

   Excessive deformation and closure of sintered, oil- impregnated bearings which subsequently require reaming.

   Rapid failure of oil-impregnated, sintered bearings whose oilways become closed during press assembly. A more extreme form of the same problem may be seen with flanged bushes.

Figure 2.20   The use of anaerobic adhesives to bond bearing, shaft and housing assemblies without recourse to press-fits. a In theory, press-fitted bearing and shafts should be correctly aligned. Where through drilling is not possible, the alignment of the holes in two separately fabricated components may be poor, either because one hole was offset or because it was not drilled at the correct angle (b). Here the shaft is distorted as the dimensional and angular discrepencies are accommodated.

The problem is avoided by using slip-fit assembly in conjunction with an anaerobic adhesive. This is shown in c where the bearing 'floats' into the correct position. The shaft is no longer bent and the assembly functions more efficiently. Generally the spatial misalignment of the bonded shaft is much less than that of the press-fitted shaft because its centre-line displacement is parallel, and therefore limited to the tolerance of the slip-fit, while in the press-fit it is angular. (See also Figures 2.14 and 2.15.)

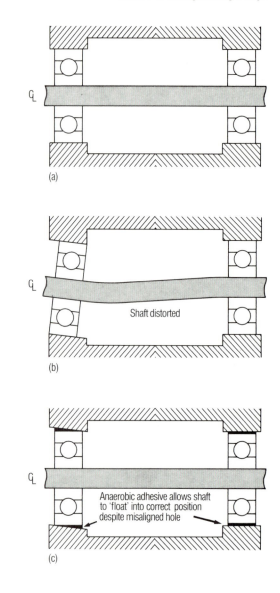

(a)

(b)

Shaft distorted

(c)

Anaerobic adhesive allows shaft to 'float' into correct position despite misaligned hole

Alignment

The angular misalignment of shafts and housings usually results from the latter being inaccurately fitted or from intermediate bearings being inaccurately positioned during their placement. Figure 2.20 illustrates the problem by showing a case where the housings had to be made separately and through-boring was not possible.

The vibration and fatigue problems likely as a result of the distorted shaft and overloaded bearings are readily apparent and, not surprisingly, premature failure often occurs in such situations.

Corrosion

It is generally agreed that no more than about 20 per cent of the design contact area in interference fits actually takes the compressive loads created by hoop stresses.

In service, the gaps between mating parts are increased by relative movement and abrasion of badly fitting components. Air and moisture enter, inducing corrosion and bringing further problems – alignment may no longer be maintained and removal may prove very difficult. Severe corrosion may cause premature bearing failure if corroded particles gain access to the raceways.

Bonding may eliminate or substantially relieve all these difficulties, but several factors still need to be considered during design:

Anaerobic adhesives can fill diametral clearances up to 0.125 mm without substantive performance loss. The gaps normally encountered in maintenance work do not normally exceed this value.

Where only one or two bushes or bearings support a shaft, it may sink through the adhesive film to lie parallel to, but below, its design position. This displacement is found to cause few problems in practice; it is certainly far less damaging than the angular displacement often caused by press fitting (see p.18 Alignment and Figures 2.14 and 2.15).

Where two or more bearings are supporting a shaft, it may 'float' before the adhesive has set but will finally rest in or very close to the design position. However, for extreme accuracy, jigs may have to be used.

Needle bearings should be fitted as specified by the manufacturer, though retention may be augmented by an adhesive.

The strength of the adhesives chosen must be appropriate for maintenance of the components.

Elevated temperatures sometimes cause significant extension of the shaft at a different rate from that of its mounting or housing. Consideration should be given to not bonding the outer race of one of the end bearings, allowing it to 'float'.

Housing weight may be reduced because bonding gives improved retention at much lower hoop stresses. This is particularly important in thin-section, light alloy housings where unwanted relative movement is sometimes difficult to control with interference fits.

Stronger grades of adhesive should be used with soft alloy or plastics housings to compensate for reduced jamming action as a result of the increased ductility of the housing (see p.25 Plastics and soft alloys).

Where alloy housings experience elevated temperatures, it may be possible to augment and standardise bearing retention if, at the operating temperature, an adhesive-supported interference fit can be devised.

The slip-fit bond technique recommended above may offer several bonuses such as a reduction in the tendency for bushes to be forced out of place, the fact that bearings will no longer need mechanical restraints and the possibility of using unground bearings. Finally, the costly process of component matching may be entirely eliminated.

Cylinder liners

Of the high-performance adhesives there is little doubt that anaerobics are uniquely capable of supplementing the retention and sealing of cylinder liners – both wet and dry. Only the strongest versions should be used and, in the most demanding circumstances, probably only the toughened variants will suffice (see Plate 17).

The fundamental engine design will influence the fit between liner and housing and, for large contact areas, tolerances may be eased, bringing stress reduction. Where contact between the liner and the block is only over the flange and its associated register, the adhesive is only used for sealing, preventing fretting corrosion and reducing surface abrasion. Even so, the adhesive will often make a positive contribution to the retention of the liner.

It is perhaps worth pointing out that it often takes twice as much force to remove interference-fitted liners whose flanges and registers have been sealed with adhesives as it does to assemble them. This is partly because the adhesive does form a bond (as distinct from its intended role as a sealant) and because of gummy deposits formed in the running engine.

Seals based on anaerobic adhesives will even function close to the combustion chambers and exhaust ports of an engine. Oxidation of the sealant and the deposition of other debris slowly replaces the adhesive with a carbonaceous, metal- and rust-impregnated material which functions effectively – if not intentionally Note that:

> Differential thermal expansion is unlikely to be a problem with similar metals.
> With different metals – particularly an aluminium block and a steel liner – differential expansion will almost certainly cause problems unless an appropriate interference fit is used.
> Only the strongest anaerobics should be used and, when flanges and registers are being sealed, toughened anaerobics should be used to maximise performance.
> Where interference (or shrink) fits are required, special slow-setting compositions should be used to prevent premature cure of the anaerobic materials. Otherwise, the seal will form too soon and in the wrong place, only to be destroyed during the final stages of assembly.
> Substantial interference fits may demand anaerobics with specially modified characteristics, which will ensure that a film will remain during the insertion of the liner and, as a consequence, the interstitial zones of the final assembly will be adequately filled.

Cams

Exploiting the design freedom of adhesives simplifies assembly of cams on to a shaft, an improvement on older methods where different metallurgical requirements of cams and shafts dictated compromises in design.

Cam shafts may experience substantial forces, often in demanding environments, and only the toughened, single-part epoxies are generally capable of sustaining such loads without premature fatigue failure. As these adhesives require positive radial clearance between the parts, jig assembly and curing are necessary and the adhesive used should be capable of capillary flow (see p.13 Toughened single-part heat-cured epoxies). Locating shoulders should be progressively stepped, each in turn acting as a seal to prevent the adhesive's egress during curing. Unless lightly loaded, the cam/shaft geometry should be designed to generate a compressive load in the adhesive during cam-follower actuation.

Fans

Typically, small fans are ideally slip-fitted with anaerobic adhesives. If disassembly is anticipated, then the medium- strength versions should be specified. Selection will also depend on the running load and engagement area. As fan size increases, and with it the weight of the components and power consumption, it will become progressively necessary to consider toughened, single-part epoxies or to introduce knurling, keys and, ultimately, splines. While knurling is not considered particularly reliable, anaerobic adhesives used with keys and splines merit consideration as the concept is technically sound and can be economically viable (see p.33 Splines, keys and set screws).

Very light plastics fans may often be bonded satisfactorily with either cyanoacrylate adhesives or toughened, cold-curing acrylics. Dismantling is not normally practicable when these adhesives are used.

Gears

Except in light mechanisms, gears should always be expected to deliver power, so simple circumferentially-bonded assemblies should be avoided. There are but few exceptions, and here only the strongest anaerobic adhesives and single-part epoxies will suffice. In general, the addition of keys, splines, flats and squared shafts are the only effective means of delivering power without causing fatigue failure (see pp.6 & 22 Fatigue and p.33 Splines, keys and set screws).

When these are used, the simpler anaerobics may be adequate, so avoiding the curing cycle of the single-part epoxies. Indeed, it may even be possible to use medium-strength grades to facilitate maintenance.

One crucial factor is the inertial loading experienced during stopping or starting. These may be much higher than the normal transmission load, and can cause premature fatigue failure of circumferentially bonded joints, which carry the load in pure shear. The use of keys and splines makes such failure most unlikely.

Impellors

The bonding of impellors to their shafts requires a similar approach to that for gears; inertial load stresses are similarly important.

Keys

See p.33 Splines, keys and set screws.

Knobs

Knobs are rarely heavily loaded nor need to be precisely positioned, and often are not removed after assembly. For these reasons, they can be easily assembled by combining slip fitting and anaerobic adhesives. Raising a burr on the side of the spindle will not hinder assembly, but it will hold the parts together until the adhesive sets.

A satisfactory method of bonding these components is to assemble them first and then apply a low-viscosity adhesive which will flow into the space between the spindle and the knob by capillary action.

Knobs are often manufactured from plastics, some of which will not bond well. In these cases the parts should be threaded and locked in place with an anaerobic adhesive, replacing over-tightening as a locking method and eliminating stress-cracking of the threaded parts. Care should be taken, however, that the anaerobic adhesive does not soften or otherwise damage the selected plastic.

Pins

While pins may be used like keys and splines to transmit loads, they are often used to locate parts in assembling precision tools. Anaerobic adhesives here offer the great advantage of avoiding the stress and damage caused by interference fits. High accuracy is easier to achieve without mechanically imposed loads and manually placed, slip-fitted and jigged pins can easily be bonded within 0.00125 mm. Experience of such techniques shows that reject rates are very considerably reduced and that the strength of the assemblies is maintained because they are not subject to the stresses generated by interference-fitted parts. Removal is also simple if the correct grade of adhesive is used.

Pipes and tubes (non-threaded)

The bonded assembly of aluminium or stainless steel pipes, tubes or sleeves, and the fitting of threaded or flanged collars to pipe-ends, is simplified by eliminating the problems of soldering, brazing and welding.

Care should be taken over any extreme environmental demands imposed by the fluids within the pipes. Contrary to popular belief, corrosive fluids excepted, water is much more damaging than many industrial chemicals; the commonly met hydrocarbons cause little if any difficulty. Generally probably only the strongest and most durable of the anaerobics and single-part epoxies are worth consideration in these applications. Durability will be extended by incorporating the following features in the design:

Maximise the engaged length of the joint and, if at all possible, make the overlap length double the pipe diameter.

Use the maximum possible sleeve or collar wall thickness.

Maximise the tube wall thickness.

Ensure that expansion loads the components in compression.

Make the bond line as thin as practicably possible – this should not exceed 0.125 mm.

Keep the imposed mechanical loads to a minimum and reduce vibration.
Prepare the surfaces properly prior to bonding.
Single-part epoxies will probably give the best performance on bonded
copper tubes for use at elevated temperatures.

This formidable list indicates quite clearly the care that needs to be taken in
such a design. Nonetheless, the implementation of known successful designs has
led to considerable savings and technical advantages. For example, unlike solder,
the single-part epoxies and anaerobic adhesives do not creep at elevated
temperatures and their performance can be better.

It is possible to bond PVC and GRP pipes successfully with toughened acrylics,
and GRP pipes can also be bonded with cold-curing, two-part epoxies. These
adhesives generally cope well with most fluids likely to be carried in such pipes.
Solvent welding may also be used to bond PVC provided that the gaps are small.

In cases other than the above, plastics pipes should only be bonded with the
approval of the pipe manufacturer. Similarly, the bonding of rubber hose requires
care and specialised knowledge.

Anaerobic adhesives are useful as a supplementary sealing medium for
compression fittings which, despite claims to the contrary, often leak. Applying
anaerobic adhesive immediately before assembling the union ensures a
satisfactory seal.

Apart from being easier to use than conventional mastics, anaerobics offer a
major benefit for sealing potable water pipes in that they are inert and do not
support microbiological activity. They will successfully seal both plain and
threaded plugs with both speed and economy.

Pulleys

The stresses in joints between pulleys and shafts are similar to those experienced
in fan and gear assemblies. However, because the belt can slip under extreme
conditions, the pulley itself is unlikely to carry the inertial loads associated with a
direct mechanical drive.

Suitable plastics pulleys may be bonded satisfactorily with cyanoacrylate
adhesives or, where large components are involved, with toughened acrylics –
many of which are unsuitable for use on small parts. The stronger anaerobic
adhesives and single-part epoxies should be used for all-metal assemblies.

Where maintenance is likely to be needed, select the weaker grades of
anaerobic adhesives (see Section 3.6).

Rotors

Electric motor rotors, particularly those with long shafts, such as in hi-fi
equipment, are best assembled using a slip-fit and anaerobic adhesives. Here,
avoiding a press fit (and its costly dimensional accuracy) ensures that the shafts are
not bent during assembly. Few problems are met with in this application, though
reduced mechanical keying of the finely ground shafts often used may demand
higher-strength grades. However, as the torque loads are usually very low, high-
strength adhesives need normally only be used if the bond area is small, or to carry

additional loads such as those due to subsequent manufacturing operations.

Thixotropic adhesives, which are not subject to capillary flow, are needed to bond laminated rotors to their shafts because the laminated rotor will remove other (Newtonian) adhesives from the joint area by capillary action, preventing a satisfactory bond.

Finally, the temperature rise in a stalled motor must be considered. If temperatures are likely to exceed 120-150°C, then heat-resisting anaerobic grades or single-part epoxies should be used.

Shafts

The advantages of adhesive bonding techniques for the assembly of shafts and associated components, discussed above in detail, can be summarised as follows:

Cracking due to excessive hoop stresses in interference-fitted parts is avoided.

Appropriate levels of accuracy are almost invariably achieved more readily and more economically.

Expensive grinding is usually not necessary.

Compared with shrink- and interference-fitted parts, bonded components extend the area engaged from about 20 per cent of that available to the maximum. Consequently the joint is sealed and fretting and associated corrosion are avoided.

Alignment (particularly angular) of bushes and bearings is better and extreme accuracy is easier to achieve.

Stress characteristics of bonded joints are invariably more uniform than those in interference-fitted assemblies.

Anaerobic adhesives – available in graded strengths – give designers a very precise, selectable performance.

Adhesives augment and standardise press or shrink-fitted assembly performance and, at the same time, seal the joint.

Adhesives ease tolerances, saving the cost of secondary machining and grinding.

Slip-fitted and bonded parts require minimal skill and equipment for assembly which, as a result, is usually quicker and invariably more economical.

Soft metals and their alloys are retained better by adhesives than by friction.

Stress-free assembly improves general dimensional accuracy and reduces distortion.

Worn or badly machined parts may be readily restored to their correct dimensions by bonding an appropriate sleeve or shim.

Additional means of transmitting load, such as interference fitting, keys, splines, knurls, pins and 'D' shafts, are only needed for assemblies that transmit power.

Gap filling and joint sealing reduce the risk of fretting, corrosion and premature failure; loads are more uniformly distributed and the easing of tolerances and substitution of turning for grinding reduces machining costs.

Sleeves

As sleeves are not generally highly loaded in shear they are well suited for retention with anaerobic compositions. Light alloy parts may need the higher-strength grades to compensate for the lower modulus of the metal (see p.25 Plastics and soft alloys).

Because anaerobic adhesives have sufficient compressive strength to support the highest load likely to be imposed by a hardened sleeve, they are an effective and economic means of extending the life of worn parts and offer a simple means of salvaging over-sized bores.

Splines, keys and set screws

When the design shear load on an adhesive joint makes overloading or fatigue failure likely (see pp. 6 & 22 Fatigue), alternative means of transmitting load from one component to the other must be examined.

Traditional solutions – splines, keys and pins – are made more effective and costs are reduced by incorporating adhesives. Almost invariably anaerobics are most suitable: even the weaker ones transmit very high compressive loads, their low shear strength makes stripping for maintenance relatively simple and there are no particular design problems.

Tolerances can be increased up to the maximum allowed by the mechanical function of the parts. The use of jigs may even allow machining errors and still ensure acceptable assembly accuracy.

Perhaps the greatest design advantage is that, as all the bearing surfaces are engaged, there is no likelihood of fretting and fatigue, so the overall dimensions of the components can be reduced, giving a possibly substantial weight reduction. This is particularly important in vehicles, where the need to reduce overall vehicle weight is combined with the goal of lower unsprung mass. A corollary of this is the marked reduction in wear and consequential backlash and noise from the non-sliding splines in vehicle transmissions.

2.3.2 Threaded components (excluding smooth co-axial assemblies)

Anaerobics are ideal for locking and sealing threaded parts and, with few exceptions (discussed separately in Section 2.3.2.9), are normally the only adhesives to consider. A variety of vibration and impact tests have shown that even medium-strength grades out-perform all conventional means of maintaining bolt tension (see Table 2.3). Frequently the test specimens fail due to metal fatigue with the nut still firmly attached to the bolt shank.

The designer's real problems are to appreciate the material's design function and to understand its nature – how it works and the various factors that affect its performance – in order to optimise its contribution.

Anaerobic compositions offer four major design features which may not be obtained collectively elsewhere:

More effective tension retention;
Elimination of interference-fitted studs and

the associated problems of casting damage;

Thread sealing; and

Corrosion suppression.

2.3.2.1 Tension retention

While popularly known as adhesives, anaerobics can more correctly be described as 'locking fluids'.

With few exceptions they are too hard and brittle to act satisfactorily as adhesives, but this very incompressibility enables them to perform well on threaded parts. Relative movement is only possible if the hardened anaerobic material is crushed between the threads. Generally this is quite impossible under the various combinations of vibration and shock loading.

The slight reduction in bolt tension illustrated in Table 2.4 is not due to any failure of the adhesive – used here as a locking medium – but to deformation of the metallic peaks, which transfers the load to the much greater surface area of the load-bearing polymer.

Anaerobic compositions and mechanical devices such as lock washers and lock nuts prevent relative movement of threaded parts in totally different ways. Any attempt to compare the performance and cost effectiveness of the two different systems on the basis of resistance to a removal torque is therefore both baseless and invidious. Particularly so since the quoted torque strength of various anaerobic materials is no more than a convenient method of ranking, not a measure of how they function. Even the weakest anaerobic adhesives can lock virtually all threaded assembly work – as shown in Table 2.3.

Table 2.3   ⅜ in.–16 tpi bolt vibration test: effect of anaerobic adhesives on vibration resistance

Surface	Condition in test	Lubricant (if any)	Treatment	Average % preload[1] left after 500 cycles
Steel	As received	Original rust proof oil	None	0[2]
Steel	Clean (solvent wash)	None	Anaerobic[3]	93.5[4]
Steel	As received	Original rust proof oil	Anaerobic[3]	91.5[4]
Zinc plate	As received (dry)	None	Anaerobic[3]	94.0[4]
Zinc plate	As received (dry)	Lube oil under head	Anaerobic[3]	88.0[4]

Notes:
1 The recommended pre-load for a bolt of this type (100 000 lb/in.^{2}) was used – the theoretical self loosening torque being 77 lb in.
2 After 180 cycles the average untreated bolt lost all its pre-load.
None of the four treated bolts lost a further significant proportion of the pre-load before their shanks broke as a result of bending fatigue. This usually occurred shortly after the 500 cycle period quoted.
3 Anaerobic used corresponded to MoD DTD 5630 – (G2) – a weak grade.
4 It is not unreasonable to assume that a substantial proportion, if not all, of the pre-load lost by the treated bolts is due to the collapse of metallic peaks as load is transferred to the polymer surface.

A wide range of adhesives is available to make it easier to dismantle large components without damage, and the relevant MoD specification (see Section 3.6) covers five adhesive strength levels, although probably no more than three would cover most industrial applications.

2.3.1.2 Removal torque

A free-running nut requires only minimal torque to move it, but once treated with an anaerobic composition that is allowed to cure, further rotation may give any one of the torque patterns of Figure 2.21. The many observations contributing to the plotting of the four typical curves of Figure 2.21 demonstrate the combined effect of various levels of compressive strength and adhesion. As will be shown later, excessive adhesion can be an embarrassment and its only justification is to augment locking performance on plain, co-axial assemblies (see pp.16, 20, 23 & 25 The general case) and particularly on soft alloys. Cured locking fluids with very low levels of adhesion function perfectly well under extreme vibration. It is perhaps worth pointing out that, although there is an initial torque required to move a 'bonded' fastener – usually known as the 'break-out' torque – it has proved notoriously difficult to measure, and accordingly the MoD test methods (see Section 3.6) are based on the 'prevailing torque' – that measured during the subsequent rotation of the fastener.

As a free-running nut is tightened down, there is a progressive increase in the torque required and in the tension in the stud or bolt. But, because of the inelastic collapse of the load-bearing peaks of the engaged surfaces, the removal torque for the fastener will normally be some 90 per cent of that required to tighten it. This residue of the pre-load is lost within some 60° of rotation during removal (see

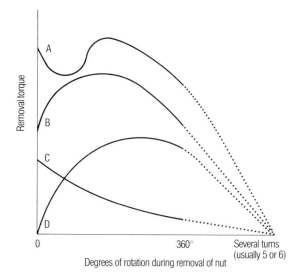

Figure 2.21   Typical torque patterns demonstrated during removal of a nut locked with an anaerobic adhesive.
A The anaerobic composition has a high level of true adhesion as well as compressive strength.
B A grade with lower adhesion and lower compressive strength than A.
C Fairly high level of adhesion coupled with a low compressive strength.
D No adhesion here, but very high compressive strength.

Note that extensive manipulation of these curves is possible: they serve only to illustrate principal features and relationships.

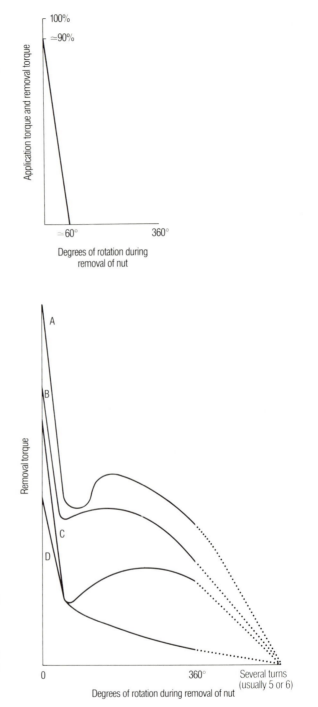

Figure 2.22   When an otherwise free-running nut is tightened down, part of the tension induced in the bolt shank is lost due to the crushing of the high-spots on the engaged threads. Thus, on removal, a reduced torque is required. This is approximately 90 per cent of the application torque and is lost over 60° of rotation at the most.

Figure 2.23   The bolt tension generates friction which is manifest in the torque required to remove the nut (Figure 2.22). This is additional to the combined effects of adhesion and crushing load induced by anaerobic adhesives (Figure 2.21). Here, with the two sets of curves combined, it may be seen that very high torques can be required to initiate movement.

Table 2.4   Typical application of the ten primary grades of the MoD anaerobic characterisation

Use	DTD no	Grade no
Screw locking and sealing	5630	2, 3
Locking and sealing nuts and small pipe unions	5631	1, 2, 3, 4
Bearing retention (removable)	5632	2
Stud locking and sealing	5633	3
Permanent high strength retention	5633	3, 4
Penetrating adhesive	5633	1
Pipe sealing (large unions)	5630	3
Gasket replacement and flange sealing	5631	4

Table 2.3 and Figure 2.22). However, with an anaerobic locking medium, the torque patterns of Figure 2.21 will apply as well on nut removal to give the composite patterns shown in Figure 2.23.

Except for permanent assemblies, pattern A of Figure 2.21 must be avoided since the combined resistance of the residual initial torque and the strength of the truly bonded joint could be sufficient to shear the stud or bolt shank on attempted removal.

In contrast to the complexities of the 'break-out' torque, the maximum prevailing torque for any grade of anaerobic adhesive will generally vary directly with the engaged length and, up to approximately 8 mm, proportionately with the diameter. Thereafter, the observed prevailing torque increases very rapidly with diameter – hence the need to use weak grades on larger diameter components, particularly pipes.

The removal torque required for the larger diameters of Figure 2.24 clearly illustrates the need for several strength levels. It is also important to ensure that a stud will not move when its associated nut is removed. Finding the correct relationship between the retention required for the nut and for the stud is relatively simple. Table 2.4 (Table 3.2 of Section 3.6, reproduced here for convenience) shows how the primary grades of the MoD anaerobic specification may be most conveniently used.

On very small-diameter fasteners with only a few threads engaged, the jamming action of even the strongest conventional anaerobic may not be

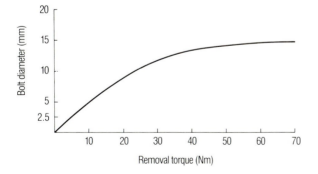

Figure 2.24   The prevailing torque required to remove a nut over a range of bolt diameters is shown by this idealised curve for a high-strength anaerobic. Although roughly proportional up to a diameter of 8 mm, the resistance to movement increases rapidly thereafter.

sufficient to prevent deliberate loosening. Only true adhesives will totally suppress movement (see Section 2.3.2.9 and p.43 Screws).

It has frequently been suggested that treated parts may be re-used without the re-application of locking compound, but this will reduce retention. This technique should only be considered after great care has been taken to test that it is truly appropriate for the design in question. Generally, light brushing to remove loose debris and the re-application of liquid anaerobic will prove much more satisfactory, and this procedure is recommended. Thorough cleaning is very seldom necessary before applying fresh adhesive after dismantling.

2.3.2.3 Sealing

Anaerobic compositions seal component threads very effectively and, while specialised materials may be required for extreme conditions (see p.28 Cylinder liners), conventional types readily meet most industrial requirements. They are more reliable and easier to use than the majority of sealants, which they readily out-perform, and consequently alleviate many severe design problems for two primary reasons. First, they have naturally stable and robust polymeric structures and second, and even more important, polymerisation creates no interstitial voids. The claim that they do not shrink is nonsense: as polyester-based resins, they shrink by as much as 10 per cent depending on type and grade. However, this does not cause leakage paths for, unlike other polymerising monomers, anaerobic compositions begin to cure where oxygen concentration is lowest – at the centre of the joint (see Figure 2.25). Thus, as polymerisation proceeds and the mass contracts, liquid is drawn inwards from the exterior fillet at the face to compensate for the lost volume, giving no apparent shrinkage and a unique sealing capacity.

Anaerobic compositions naturally have a very high compressive strength and

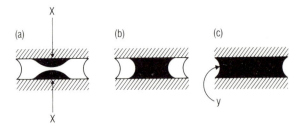

Figure 2.25   Anaerobic compositions polymerise progressively from the centre of a joint outwards. The catalytic process starts at X where the oxygen replacement (by diffusion) is minimal.

The process continues until all the material within the joint (which may be threaded) is solid – though a thin film (y) of unpolymerised and still liquid material may remain on the exterior (meniscus) face. This is maintained by its ready access to atmospheric oxygen.

plasticisers must be added to create the weaker grades required for maintenance. Plasticisers weaken the polymeric structure, not only reducing mechanical strength but also lowering the environmental resistance of the cured material. However, even the weakest of the commonly used grades make excellent pipe sealing media which will cope readily with the conditions found in central heating systems.

Corrosive media and the more searching fluids test anaerobic compositions to their limit. Careful evaluation is needed where a design must cater for these materials (see Appendix 2).

Finally, if the unions of gas-bearing pipes are liable to experience impact damage or very low temperatures then the compliant rubber toughened anaerobic compositions should be used, as these are not prone to forming hairline cracks.

2.3.2.4 Corrosion suppression

Anaerobic compositions suppress inter-component thread corrosion very effectively by preventing the free passage of oxygen and water – both essential for the process of corrosion. Thus, in bad environments such as coal mines, they are frequently used to make maintenance possible simply by preventing formation of corrosion products which might otherwise irretrievably lock threaded components.

2.3.2.5 Factors affecting performance

Remember that the only satisfactory way of ranking anaerobic thread-locking media is to measure their prevailing torque by Method D of MoD, DTD 5628, which does not assess their actual anti-vibration capacity (see Section 2.3.2.1 and Table 2.3).

But prevailing torque will be affected by any variation in the standard conditions required by Method D. For example, the prevailing torque will be increased by:

Surface area;
Surface roughness;
Hardness; and
Reduction in temperature.
And will be decreased by:
The presence of oil;
A reduction in surface hardness (caused by plating or the use of alloys or plastics); and
Increase in temperature.

Thus any attempt to rank and compare anaerobic media not using the standard method is very likely to lead to an erroneous conclusion. Again, it is worth emphasising that even the weakest of these materials will resist severe vibration, and it is usually the combined dictates of maintenance and environmental exposure that demand a variety of 'strength' levels.

2.3.2.6 Microencapsulation

Threaded fasteners are occasionally supplied pre-coated with a locking medium in the form of a band of microencapsulated epoxy or anaerobic adhesive around the appropriate section of the screw thread. This ensures that the adhesive application is not overlooked. While microencapsulation is attractive and useful, its chequered history indicates limited potential; any one of the following major drawbacks may be encountered in practice:

Loss of locking medium due to storage damage;

A progressive loss of reactivity leading ultimately to a complete failure to cure (unlikely in epoxy-based systems);

Premature polymerisation;

Substantial and variable increase in the torque required to place and tension the fastener;

Little or no contribution to the initial torque required to remove the fastener, should this be required;

Substantial variations in prevailing torque;

Very poor sealing characteristics.

These observations should be carefully considered before choosing between liquid and encapsulated systems.

2.3.2.7 Post-assembly thread locking and sealing

Occasionally, it is convenient to apply the thread locking medium after assembly. Low-viscosity anaerobic compositions (10–15 mm^2 s^{-1}), ideal for these situations, will generally be readily drawn into the thread by capillary action. If the joint is to be sealed as well as locked, then appropriate techniques should be developed (see Figure 2.26).

Similarly, non-threaded assemblies as well as porous castings and welds can be sealed.

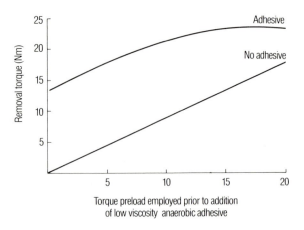

Figure 2.26   Post-assembly thread locking. A high-strength, low viscosity anaerobic adhesive has been used to lock an M8 nut and bolt assembly. The adhesive was applied after the nut was tightened. Actual results obtained from a material to DTD 5633/G1 with a performance characteristic similar to pattern B in Figure 2.21.

2.3.2.8 Thread lubrication and the torque/tension relationship

Automatic nut tightening equipment is used to ensure that the required clamping load is achieved; where this is not available, the traditional method of applying the appropriate torque for the required bolt tension must be used. Because far more energy goes in overcoming friction than in stretching the bolt, any change in thread lubrication has a major effect on the clamping load achieved. For standard fasteners tables are available giving figures for under-head and thread friction. The figures normally relate to a standard oiled fastener and are claimed to relate tightening torque to clamping load.

Problems can arise if thread friction is unknown – for example when the bolt has been cleaned or coated with an aerobic adhesive. In an extreme situation the use of an adhesive with a sufficiently different lubricating effect to machine oil could result in a given torque generating insufficient clamping force. The clamped parts might move under severe vibration even though the fastener itself would not loosen.

In Figure 2.27, the torque/tension relationship for standard anaerobic compositions on oiled and degreased bolts shows that the reduction in tension only appears with adhesive on a degreased fastener.

Specially developed 'lubricated' grades of anaerobic adhesive counter this and the effect of one such is shown in Figure 2.28. Coating both oiled and degreased bolts with this particular composition gives the same sort of torque/tension relationship as an oiled 'as received' fastener.

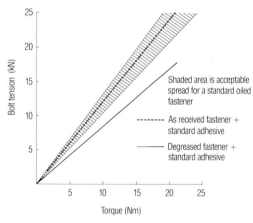

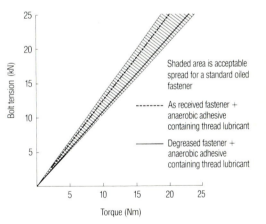

Figure 2.27   When an M10 bolt is treated with a conventional anaerobic, the torque/tension relationship falls within the accepted limits provided that the bolt has not been degreased prior to adhesive application. Cleaned bolts fall outside the allowed limits.

Figure 2.28   By comparison with Figure 2.27, special lubricants contained in the modified anaerobic adhesive lift the torque/tension relationship back into the accepted zone, even when used on degreased bolts.

It is perhaps worth emphasising that a nut locked with an anaerobic composition will function effectively without cleaning it or the corresponding fastener. But, although vibration is countered by the hard polymer, preventing relative movement, the actual removal torque will be less than if it had it been cleaned. It is essential not to confuse the function of the material – a hard incompressible solid that counters vibration – with what is, in fact, a test method providing a ranking order under idealised conditions (see Section 2.3.2.5).

2.3.2.9 The use of true adhesives – materials other than anaerobic compositions

As pointed out in Section 1.1, most anaerobic compositions are not true adhesives and, although very stable, they are not currently as robust or durable as the heat-cured, toughened, single-part epoxies. These epoxies are therefore used to replace solder on brass components when only a few threads are available for engagement. The softness of the alloy coupled with the low engagement area militate against anaerobics, especially when it is intended to use these at high temperatures.

At the other extreme, cyanoacrylates may be used to prevent very small threaded parts from moving (see Section 2.3.2.1), but are not suitable for hot, wet conditions.

It is very difficult to render the smaller screw sizes absolutely immovable with anaerobics. While they will prevent vibration loosening, they cannot necessarily be relied on to prevent removal where permanent assembly and total security is required. In such circumstances cyanoacrylates offer the advantage of literally bonding the parts together, as well as giving a hard mechanical wedge to lock them.

2.3.2.10 Specific applications

The stud

A great bonus for the designer is that loose-fitting (Class 2) studs can replace the more expensive interference-fitted version so often needed when differential thermal expansion occurs in service. Often, despite necessarily increased component dimensions to prevent their cracking during interference fitting, components may distort and the possibility of a stud stripping or shearing is not avoided. The use of adhesives saves weight, reduces the risk of component damage and simplifies maintenance and the whole process of manufacture and assembly.

As neither special skills nor equipment are required, studs may be fitted anywhere. Selective fitting is avoided as also is the associated use of taps and dies. Absence of thread run-out, no longer required as a locking aid, brings additional cost savings and greater design freedom. Further design flexibility comes from the sealing action of the anaerobic locking medium; blind holes are no longer necessary to avoid leakage.

It is normal practice to secure studs that do not have to be removed for maintenance with the strongest grade of adhesive so that the stud will never be unscrewed when the corresponding nut is removed. However, if stud removal may be necessary, then the grade chosen will depend upon the engaged length of the stud and the depth of the nut.

Normally this relationship allows the same grade to be used for both, but as a safety precaution it is better to use a lower-strength material for locking the nut than the stud, to ensure that the stud will not be disturbed during removal of the nut.

When a stud is inserted into a blind hole, the escaping air may blow some of the still liquid adhesive out of the thread. This must be prevented to give complete thread filling in critical applications by machining a slot or flat on the side of the stud.

The nut

Locking a nut with an anaerobic adhesive is far more effective and economical than using mechanical lock nuts, which are invariably more likely to be loosened by vibration and shock. They are also more expensive and do not prevent corrosion (see Sections 2.3.2, 2.3.2.1 and 2.3.2.4).

Nuts locked with the correct grade of anaerobic adhesive can readily be removed with conventional tools and, where equipment is used in very severe environmental conditions, anaerobic adhesives are frequently employed to prevent corrosion.

Weaker anaerobics should be used with soft alloy nuts and other threaded fasteners to prevent thread stripping during disassembly. Soft alloy and plastics fasteners exhibit a lower removal torque than their steel equivalents, because less hoop stress is generated as the adhesive is crushed during removal. Despite this apparently lower performance (remember removal is only a test method ranking) the parts will still be effectively locked and resistant to vibration.

A distinction needs to be made here with the observations of Section 2.3.1.3 (p.25), recommending use of the higher-strength anaerobic compositions – preferably those with true adhesive characteristics – to ensure retention of fitted non-threaded parts fabricated from soft alloys and plastics.

Screws

Just as with nut locking, anaerobic compositions used with screws give greater security, reduced corrosion and reduced likelihood of damage while still allowing parts to be dismantled with conventional tools. These features are particularly useful when no axial tension can be generated in the screw for locking.

Note that, for screws engaged over the greater part of their length, only an appropriately weak grade will allow easy removal for maintenance. Furthermore, such a weak material will also allow adjustment and retention – but this technique should only be employed on 'safety-critical' items after very careful evaluation. Occasionally, a screw needs to be made quite immovable and, for this, only the truly adhesive versions of the anaerobic family or some other suitable permanent adhesive should be used. A cyanoacrylate is a practical alternative and, although epoxies may also be used, they are less convenient because they are often viscous and need either mixing or heating for curing. The absolute immobilisation of screws with diameters less than 3-4 mm and only a few threads engaged can prove particularly difficult and careful evaluation is required.

Soft alloy and plastics screws may be locked with anaerobics but the softness of the material will reduce the prevailing torque. If a plastics screw is being placed in a plastics part then a secondary catalyst will be needed in addition to the anaerobic composition. This will normally be an aerosol-based primer or, for large batch quantities, total immersion in the primer may be practicable.

Pipe unions

Although they are far from being a general' panacea, anaerobics offer greater design freedom and security than any other means of locking and sealing pipe unions. Not only will they lock the assemblies to prevent undesirable relative movement, but they will also seal the engaged threads. Although they may not withstand environmental extremes – such as very hot steam – they can deal very successfully with the majority of conditions. This is because they are solvent-free, and complete conversion from liquid or paste to a strong solid involves no shrinkage within the joint and hence no gaps, as discussed in Section 2.3.2.3. As the chemically robust and strong solid formed can be neither readily moved nor destroyed, they do not extrude, degenerate or simply leak as do conventional mastics – particularly those incorporating hemp.

A number of consequent advantages can be exploited by the designer:

Unions based on either tapered or parallel thread forms may be sealed without being wrenched tight.

Generally, fittings may be positioned accurately – both axially and angularly – without the over- or under-tightening problems of conventional mastics. It should be noted that normally at least three complete threads should be engaged and sealed to provide mechanical stability and an appropriate surface area for sealing.

There is much less reliance on powerful tools, and in some instances their use may be avoided altogether. This is particularly valuable for site work since this often has to be carried out under bad conditions and in confined spaces.

Extremely high pressures may be borne which, according to design, can be as high as 1000 bar.

Destructive materials such as the hydraulic fluids based on phosphate esters may be contained (see Appendix 2).

The high-viscosity/thixotropic grades will ensure 'instant' sealing up to approximately 30 bar on standard BS pipe threads.

Conversely, the low-viscosity versions are completely suitable for the finely threaded small-bore pipes used in gas appliances (see BS 5292: Section 7).

Anaerobic compositions are not biodegradable – unlike hemp and oil-based mastics – and so are recommended by the National Water Council and the

Department of the Environment for situations which involve contact with potable water.

As well as these conceptual advantages, very real production benefits stem from anaerobics, which are quicker and easier to use than the conventional sealing media. Coupled with their vastly superior reliability, this ensures much less re-work – a particularly acute problem with fibrous packing or tape, where not only is correct alignment difficult, but also fibre or tape shards entering the pipe bore may subsequently lead to equipment failure elsewhere in the system.

Temperature

The maximum working range of the whole class is approximately -50°C to +200°C. Above this temperature, chemical degeneration will inevitably take place, although the oxidised polymer may continue to function as a seal supported by the deposition of naturally occurring mineral and carbonaceous deposits.

Extreme environments

Anaerobic adhesives should not be used on pipes carrying live steam without very careful consideration, and generally this is not recommended.

At the lower end of the temperature range, the hardness and consequent brittleness of anaerobics becomes a disadvantage, promoting microcrack formation when subjected to mechanical shock. With gases, the results may range from an inconsequential leak to a potentially dangerous situation. Where such damage is likely, the highly flexible, rubber-toughened, anaerobic versions should be used, for these appear to cope well with such situations.

Anaerobic compositions are very robust, but are subject to attack and degradation by the more corrosive chemicals. The higher the temperature and the greater the concentration, the more severe the attack. Nonetheless, out of the 1000 or so common industrial chemicals only about 10 per cent – most of them fairly obvious – are likely to cause difficulty (see Appendix 2).

The most durable anaerobic compositions, usually the toughened unplasticised versions, should be specified for very aggressive conditions. However, this may make disassembly without damage extremely difficult or impossible unless parts can be heated to about 200°C.

Application technique

For pipe unions and plugs, thixotropic versions are best and should be applied to the male thread. The use of excessive quantities of a low-viscosity type may result in liquid adhesive entering the pipes. Fortunately, except with gross con-tamination, the liquid material will normally disperse harmlessly.

2.3.3 The lap joint and its variants

Lap joints and their variants are very common, particularly in the assembly of large load-bearing structural components. Their effectiveness depends on the materials involved, the geometry of the whole assembly, and the scale and disposition of the forces to be borne.

Consider the joint formed around the axis x-x of Figure 2.29a. As load increases, the amount of deformation of the assembly will depend upon the stiffness of the adherends. Ultimately, in the extreme position in Figure 2.29b, the opposing forces will have become aligned along the same plane due to the deformation of the original geometry of the joint around x-x. Distortion, particularly severe at the end of the joint, induces secondary tensile forces within the adhesive in that area. These forces (Figure 2.29c) can rise to very high levels which may cause the failure of the adhesive. Careful observation shows that in practice simple lap joints destroyed by shear loading actually fail because the bonded faces peel apart. With very powerful adhesives, peeling loads become severe enough to cause inelastic deformation of the adherends, which remain permanently bent at the end of the test. This provides a qualitative insight into the orders of load borne by the adhesive immediately prior to failure. More specifically, the damaging effect of a peeling load can be gauged from the fact that a structural adhesive with a shear strength of the order of 35 MPa under standard conditions will fail under the influence of a 25 kg load when assessed by means of a typical peel test – for example, the 'T' peel.

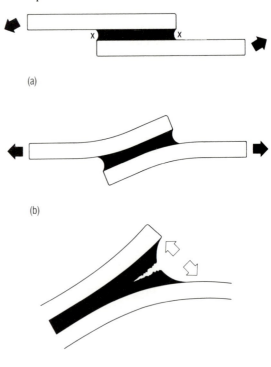

Figure 2.29   Development of secondary destructive tensile forces at the edge of a joint. a Initial loading is not parallel to the joint axis because of asymmetric position relative to the tensometer grips. b As the load increases, the assembly becomes progressively distorted as it is forced into a symmetrical configuration. c Rupture takes place as the secondary tensile forces generated in the face of the adhesive at the edge of the joint exceed either its cohesive strength (shown here) or its adhesion to the adherend's surface. Once initiated, failure may be catastrophic – particularly with brittle adhesives.

(a)

(b)

(c)

The shear stress pattern in the joint, illustrated in Figure 2.29, depends on many factors. In a stiff adhesive the load is carried by areas closer to the edges of the joint, whereas a more compliant material will transfer load towards the inner areas (see Section 2.2.4 and Figures 2.5-2.9 inclusive). Where a stiff adhesive is used, the extension of joint length beyond, say, 10 mm, will offer only a marginal increase in overall performance because once this length is exceeded, the inner areas will not be loaded. By comparison, extra joint length brings greater benefit with a compliant adhesive, but even here the contribution will become disproportionately small once joint length exceeds about 15 mm.

As might be expected, the effect depends also on the elasticity of the adherends, their thickness and that of the adhesive layer.

If peeling forces are counteracted by constraining the joint with lubricated clamps, this type of failure is not observed. Indeed, when high-performance adhesives are used, the shear loading of restrained joints usually tears the adherends – even metal ones – and the joint itself remains intact.

Mechanical restraints are usually not appropriate in practical working joints, which should therefore be designed so that, wherever possible, both elastic and inelastic deformation of the adherend load the adhesive in compression.

Of course, this is not always possible, and inclined loads that tend to peel and cleave one component from another must be countered. A number of practicable means of doing this are reviewed in Section 2.3.3.7.

By comparison with their poor performance in peel and cleavage, adhesives can support compression loads extremely well. Indeed, the stronger materials are not destroyed until these exceed 350 MPa. Thus, wherever possible, a structural joint should always be designed to distribute imposed loads within the adhesive layer as a combination of shear and compressive forces.

2.3.3.1 The simple lap joint

This basic joint configuration can cope very adequately with a wide variety of adherend types and adhesives. While the major limitation is poor peel and cleavage resistance, it performs well with appropriate materials – for example, flexible, compliant rubber bonded to a second, stiffer adherend. The stress induced at the joint's edge by distortion will be dissipated over a greater area than usual by the bending of the rubber and the more closely matching cohesive strengths of the adhesive and the rubbery material – compared with those of an adhesive and metal – could well mean that the joint will ultimately fail only because the rubber tears at extreme load.

Designs deliberately incorporating flexible rubbery inserts between two stiff adherends are well known. For example, cyanoacrylate adhesives are successfully used to bond spectacle lenses to frames through an intermediate rubber layer. Without the rubber to dissipate peel and cleavage loads, the joint between the lens and the metal frame would be readily over-stressed, resulting in premature failure.

The simple lap joint may not perform well in load-bearing structures when other considerations dictate the use of untoughened adhesives, because the distortion illustrated in Figure 2.29 may cause brittle failure at quite low load levels, making the joint collapse catastrophically.

A number of techniques (see Figure 2.30) have been devised to prevent the premature catastrophic collapse of joints using brittle adhesives. Regrettably, although often useful, such modifications are either expensive or difficult to implement. It is better to use a toughened adhesive if at all possible, preferably in conjunction with the more refined versions of the lap joint discussed below.

2.3.3.2 The rebated (joggled) lap joint

This excellent variant of the simple lap joint is effective for bonding metal to metal (particularly sheet metal) and metal to plastics. Unless gross distortion causes peel and cleavage, normal structural loads induce compression forces in the critical region indicated in Figure 2.31, ensuring very robust joints – particularly with toughened adhesives.

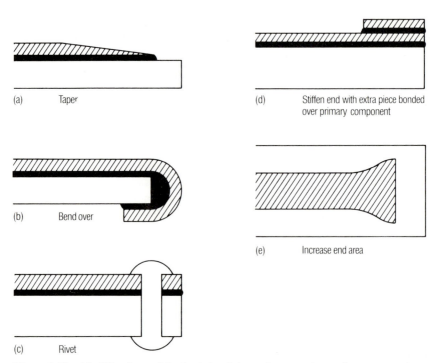

(a)   Taper

(b)   Bend over

(c)   Rivet

(d)   Stiffen end with extra piece bonded over primary component

(e)   Increase end area

Figure 2.30   Modifications of the basic lap joint to improve its performance and prevent its end from lifting. They are particularly useful when brittle adhesives have to be used. However they are, in the main, expensive and cumbersome and may often be avoided if toughened adhesives can be used.

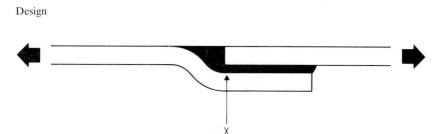

Figure 2.31   The rebated (joggled) lap joint. When shear loaded, the correct geometric alignment of the adherends inhibits the generation of tensile forces, normal to the plane of the joint, (at X) which may cleave a simple lap joint prematurely (see Figure 2.29b). This type of joint is particularly useful for sheet metal assembly.

Step   Double step

Figure 2.32   The stepped lap joint. The two forms of this joint are commonly used, particularly for wooden and moulded plastic joints, where shaping and forming costs are low. However, their use on metal structures is limited by machining costs unless extrusions can be used (see Figure 2.35).

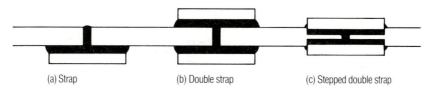

(a) Strap   (b) Double strap   (c) Stepped double strap

Figure 2.33   The strapped lap joint. a Simple and cheap joint but poor if there are any peel and cleavage forces generated by distortion. b Simple and cheap joint but it requires access to both sides of a structure. It can give a very robust performance. c An even more effective form but one which can involve considerable fabrication costs if metal machining is required and extrusions cannot be employed (see Figure 2.35).

2.3.3.3 The stepped lap joint

The single and double-stepped lap joint in Figure 2.32 is a further refinement of the rebated or joggled lap. The bonded vertical faces help to reduce the tendency to fail either by peel or cleavage.

This joint is compact and can be accurately made, though it can be expensive – particularly in metal. Nonetheless, it is popular in the manufacture of injection mouldings as it allows the accurate placement of the plastics parts. Such joints may be also made at virtually no extra cost in GRP components, using any of the shaping methods for such parts.

2.3.3.4 The strapped lap joint

Variants of the strapped lap joint and the related double strapped lap joint are illustrated in Figure 2.33. The single strap joint is suitable only for light loads, since it is clearly prone to premature failure if subject to peel and cleavage forces.

Figure 2.34   The chamfered, or tapered, double strap lap joint. This combines the best features of Figures 2.30a and 2.33b to give an extremely robust joint. Costs need not be high if the straps can be extruded (see Figure 2.35).

The double-strap and double-stepped versions overcome this susceptibility. The stepped variant may be suitable for a range of plastics components, though too expensive for most metal-based structures.

The double strap is the preferred form to give substantial load- bearing capacity, which may be enhanced by attention to joint geometry – especially important where untoughened adhesives must be used. Chamfering the strap to give the tapered double strap of Figure 2.34 improves joint strength: the greater compliance of the tapered tip lowers stress concentration of the edge of the joint, reducing any tendency for peel and cleavage failure. With aluminium, the tapered double strap can exploit extrusions with great advantage – particularly in association with the assembly of sheet metal and plastics. The examples shown in Figure 2.35 are clearly not the only possible profiles and arrangements.

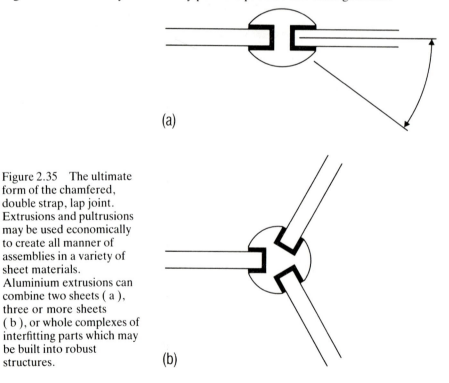

(a)

Figure 2.35   The ultimate form of the chamfered, double strap, lap joint. Extrusions and pultrusions may be used economically to create all manner of assemblies in a variety of sheet materials. Aluminium extrusions can combine two sheets (a), three or more sheets (b), or whole complexes of interfitting parts which may be built into robust structures.

(b)

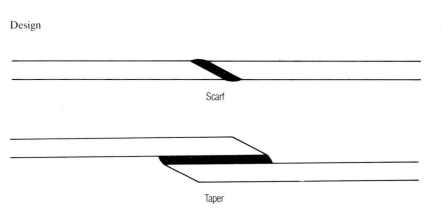

Scarf

Taper

Figure 2.36   The scarf and tapered lap joints. While effective in appropriate designs, these joints are always more expensive than simple lap joints – especially when formed from metal. Often the scarf will be quite impracticable.

2.3.3.5 The scarf and tapered lap

Both the scarf and tapered lap joints in Figure 2.36 can perform well – particularly the scarf joint. However, the scarf is not very practical in thin sheet metal and both joints can be expensive to fabricate unless wood or plastics are used.

2.3.3.6 The butt joint

The butt joint is generally regarded as an extremely poor form of lap joint; Figure 2.37 shows three versions. However, its success depends both on its geometry and the physical characteristics of the adherends. Consider Figure 2.37a. If the material involved is a rubber cord – such as an 'O' ring – then a strong, robust joint is formed because all the stresses induced by loading the joint are dispersed within the rubber. Typically, stretched beyond its limit, the rubber will tear leaving the

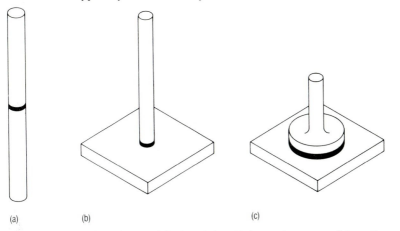

(a)   (b)   (c)

Figure 2.37   Various forms of the butt joint. Unless at least one of the adherends is highly compliant, assemblies a and b will fail readily as even minor misalignment will induce very high peel and cleavage forces which will rupture the joint (see Figure 2.29c). However, a reduction in height and an increase in diameter (c) will stabilise the joint.

joint itself intact. The other extreme, bonding two non-compliant metal rods end to end, is courting disaster. The slightest load component at right angles to the axis of the assembly will create unsupportable cleavage forces on the adhesive and the joint will fail. The geometry of the joint is fundamentally important. Consider Figure 2.37b. As rod length decreases and diameter increases, the position shown in Figure 2.38a is eventually reached, where very sharply reduced leverage effects allow considerable loads to be borne. Some fasteners exploit a very practical variation of this design by coupling an extended base with a relatively narrow shank which will break before sufficiently large cleavage forces are generated to rupture the bonded base.

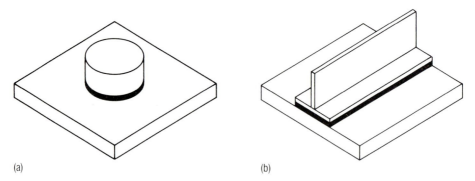

(a)   (b)

Figure 2.38   The developed form of the butt joint. Ultimately the lowering and widening of the column in Figure 2.37b results in the truncated form above (a). This structure is so inflexible that an extremely strong joint is formed. Derivatives of this are used to attached threaded fasteners or may serve as stiffners in the form shown in b , which is not necessarily recognised as a butt joint – though its origins are clear.

2.3.3.7 Practical structures

Figure 2.39 highlights correct and incorrect methods of designing or assembling a variety of bonded structures. The key lesson in every case is to load the joint with a combination of compression and shear forces and to avoid destructive peel and cleavage loads.

Combination joints using either screws or rivets may also be used and sheet metals can be successfully welded through a layer of suitable adhesive.

Screws, rivets or welding are useful in holding components together while the adhesive cures and, where peel or cleavage forces cannot be avoided, they can be used to counteract such loads. Although the adhesive effect may well be secondary, it will still make a major contribution to the overall stiffness of the structure.

Real joints must allow cost-effective assembly. It is usually better to design joints so that the parts are pressed down upon the adhesive rather than slid into position. 'Placement on' rather than 'sliding over' ensures that the adhesive is not pushed out of place (see Figure 2.40).

Large areas may require considerable pressure to squeeze out the adhesive.

To avoid air entrapment, one component should be lowered slowly over the other, preferably starting from one end. If possible, pressure should be applied and maintained progressively over the surface.

The design must also allow the adhesive to be positioned in an appropriate pattern. Stripes and crosses are preferred so that placing the components together does not entrap air as is possible with a closed-loop pattern.

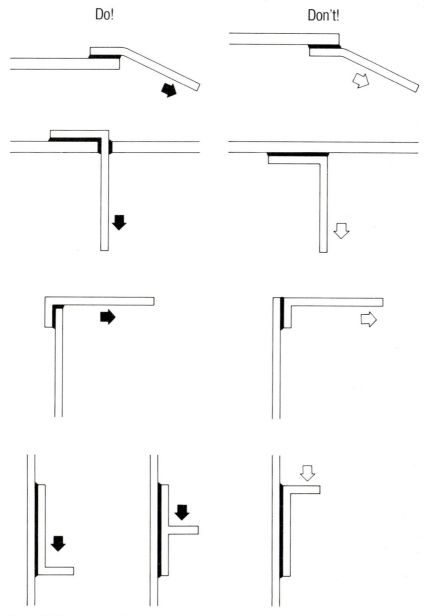

Figure 2.39   Acceptable and unacceptable practice in joint design.

Figure 2.39   (*continued*)

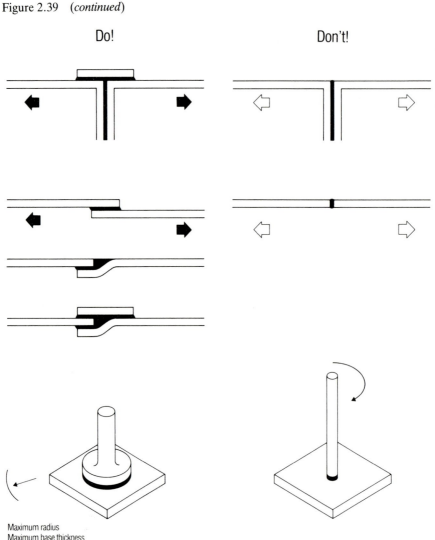

Maximum radius
Maximum base thickness
Minimum pillar width

Adhesives may be more accurately placed if they are deposited in a shallow groove, or if there is a convenient ledge nearby that can be used to guide the application gun.

Finally, the excess adhesive squeezed from a joint can contribute to its overall stability by reducing stress concentration at the edge, so extruded adhesive should only be removed for overriding practical or aesthetic reasons.

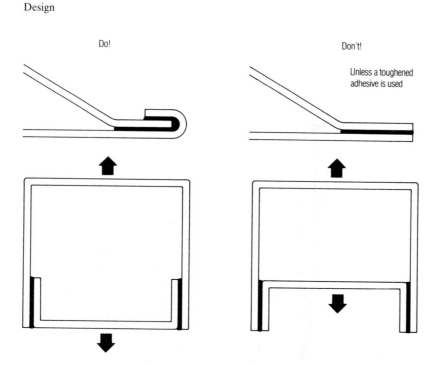

Figure 2.39 (*continued*)   This structure is based on welding practice and does not favour the adhesive. By reversing the closing panel, the load in the joint becomes compressive and the left-hand box is 3-4 times stronger than that on the right.

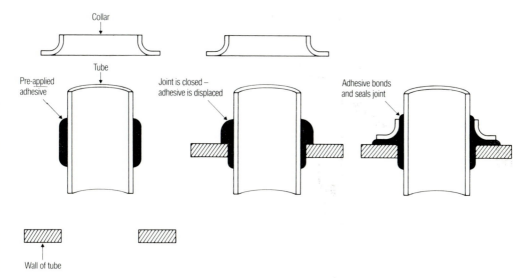

Figure 2.40   Design of the T junction in tube assembly, showing correct adhesive distribution and air exclusion achieved by pressing the collar into position.

2.4 Sealing

Sometimes bonded joints must form a positive seal. Some adhesives readily lend themselves to this dual role while others, for various reasons, do not. The selection procedure given in Section 5.2 highlights likely adhesives.

The more robust materials – primarily the toughened anaerobics – are being increasingly used as gasketting media, contributing to the current trend towards the replacement of conventional gaskets by solvent-free sealants. Gaskets, even with the help of conventional sealing compounds, cannot overcome many production problems including the following:

Gaskets must conform to all surface irregularities and block these imperfections to effect a seal. This is very difficult, even with the softest materials, and a compromise is the combination of a firm gasket and a non-hardening fluid or paste. Despite considerable ingenuity, leaks can still occur because the secondary sealant may be displaced or the gasket lose its resilience, destroying the clamping pressure.

Such clamping pressure loss may bring about relative movement, induced by both tensile and torsional loads, which may lead to premature seal failure.

The variable compression of traditional gaskets makes overall assembly tolerances difficult to achieve and maintain.

Although they are no panacea, the solventless anaerobics ease the situation considerably, with the following specific advantages:

Their excellent wetting property readily fills all surface imperfections.

Total conversion of liquid sealant to solid minimises void formation.

Metal-to-metal contact combined with the high compressive strength of the adhesive/sealant ensures maintainable dimensional stability.

If necessary, the truly adhesive versions of the anaerobic compositions can be used to reduce any likelihood of tensile and torsional loads causing movement and consequential leaks.

The chemical stability of the more durable compositions makes them less prone to solvent attack and other forms of degeneration.

The major limitations of all current materials are an inability to cope reliably with temperatures in excess of 200°C and a tendency for thinner-gauge flanges to flex, where traditional materials may cope much better. However, it appears that some of the more flexible toughened anaerobics may perform well even when subject to considerable distortion.

2.5 The other manufacturing processes

Joint design requires consideration of:

Component material;
Shaping and finishing;
Material preparation;

Geometry and physical dimensions;
Adhesive selection;
Adhesive placement and clean-up;
Curing process;
Curing time;
Temporary jigs, clamps etc, for retaining;
Component position during the curing process – vertical or horizontal; and
Inspection and quality control.

Each of these can be complex. Adhesive placement may be as complicated as an independently monitored automated metering, mixing and dispensing system fitted with a pot-life guard – but it could be as simple as manual dispensing from a purpose-made bottle.

However, all of these topics are well recognised as vital to the manufacturing process and are, therefore, less likely to be overlooked than their interaction and relationship with other stages of production.

With so many differing manufacturing processes, it is quite impossible to discuss detail, but a few examples illustrate some of the potential hazards to be avoided:

Will any heating process – say, associated with painting or welding – degrade the adhesive or allow components to separate due to temporary softening of the bond?

Will inadvertent jolting or loading cause relative movement of the components before the adhesive has cured?

Will subsequent rivetting, clenching or other impacting processes cause joint fracture?

Might the adhesive contaminate other areas or surfaces to the detriment of subsequent processes, such as painting?

If electrical conductivity is important, as in electrostatic spraying, for example, will this be impaired by the adhesive?

Will surface preparation cause problems such as the handling of toxic wastes or the corrosion of freshly cleaned steel before further protective coatings can be applied?

Will unexpected changes in component temperature (due to outside storage, for example) cause unacceptable variations in the adhesive's cure time?

Although the list could be continued almost indefinitely, production lines employing bonding processes do run smoothly year in and year out once established.

Any designer unfamiliar with the latest developments in bonding technology, however, must operate closely with the manufacturer of the intended adhesive in thorough evaluation, preparation and pre-production trials. This should result in the speedy selection and implementation of a bonding system that will prove technically successful and cost effective.

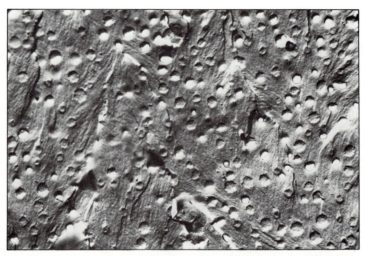

Plate 1   The internal structure of a typical toughened adhesive is revealed by the electron microscope. The rounded zones are the rubbery domains which give adhesives of this type their characteristic resistance to brittle failure. Each domain is approximately one micron in diameter.

Plate 2   Anaerobic adhesive is used to retain the oil-impregnated bush shown at the centre. Its use here, and for the retension of the bush supporting the end of the drive shaft, enables the gears to take up the correct spatial alignment without recourse to rigorous accuracy in their respective housings. Material: Permabond A115.

Plate 3a   This gear is part
of the drive mechanism of a
steel rolling mill. After
many years of use, the shaft
cracked and could not be
repaired by welding.
Instead the fractured end
was cut away and a new
driving collar fitted. The
shaft and collar were
bonded together.

Plate 3b   Application of
the anaerobic adhesive
prior to insertion. Several
years after being repaired
the bonded components are
still giving daily service.
Material: Casco ML Grade
MLF118. (Photograph
courtesy Delo GMBH.)

Plate 4   Anaerobic adhesives and sealants are used extensively in the assembly of engines built by Lotus Cars. The one illustrated is the turbo-charged 2.2 litre engine. (Photograph by Focalpoint, courtesy Lotus Cars Ltd.)

Plate 5   A thixotropic anaerobic sealing compound being applied to the lower face of the 2.2 litre aluminium engine block. The liquid sealant cures spontaneously to form a flexible, robust seal between the block and the steel baffle place which separates it from the aluminium sump. The sealant is also used on the face between the baffle plate and the sump. Being anaerobic in nature the liquid sealant cures spontaneously when the parts are assembled and access to atmospheric oxygen is denied. The flexible seal which is produced is not only resistant to hot engine oil but also accommodates high frequency vibration and the differential expansion of the various components. Material: Permabond A136. (Photograph by Focalpoint, courtesy Lotus Cars Ltd.)

Plate 6   The flywheel of the 2.2 litre Turbo is retained with six bolts. These are treated with an anaerobic adhesive to prevent loosening. Considerable stresses are involved for the flywheel can rotate up to 7000 rpm. Not only must the anaerobic adhesive prevent the loss of tension within the bolts but it must also seal the thread in order to prevent oil loss. Material: Permabond A115. (Photography by Focalpoint, courtesy Lotus Cars Ltd.)

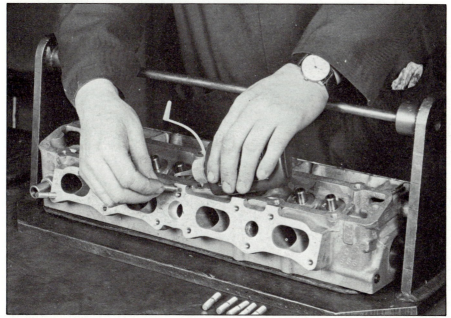

Plate 7   The studs used to secure the inlet manifold to the cylinder head are retained in position with an anaerobic adhesive. Although of high strength, the grade chosen will allow maintenance and ready removal of the studs should this be necessary. Because the studs are a loose fit – and hand fitted – and retention is ensured by the adhesive, no stresses are generated in the aluminium cylinder head. This means that the sections employed may be of minimal thickness as there is no danger of cracks being initiated by over-tightened studs. The studs of the exhaust manifold (rear face of head) are treated in a similar manner. Note that, when the Turbo modified engine is being built these studs are fabricated from stainless steel. Material: Permabond A115. (Photograph by Focalpoint, courtesy Lotus Cars Ltd.)

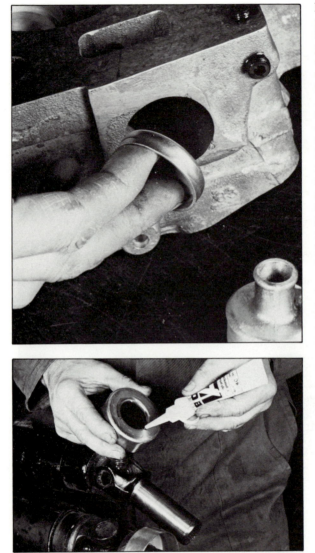

Plate 8   Three types of plug may be seen in this illustration. A simple cup plug is being fitted to the core plug hole in the front end of the casting which forms the cylinder head. Advantage is taken of the core plug hole when the engine is adapted to take a turbo-charger. The extra heat generated requires an increased water flow to the cylinder head and this is achieved by bonding the special plug shown in the foreground into the hole normally filled by a cup plug. Both are bonded with a high strength adhesive (material: Permabond A138). A much weaker grade – which allows maintenance should this ever be required – is used to retain and seal the tapered and threaded plug which seals the end of the oil gallery supplying the cam housing. Material: Permabond A121. (Photograph by Focalpoint, courtesy Lotus Cars Ltd.)

Plate 9   A steel collar – which forms a protective seal – is fitted over the end of the propellor shaft and bonded in position using an anaerobic adhesive. This is placed on the face of the inner radius where the metal is formed to give an area of axial engagement on the outer face of the propellor shaft. No other means of retaining the collar is employed. Material: Permabond A140. (Photograph by Focalpoint, courtesy Lotus Cars Ltd.)

Plate 10a   Prior to fitting the piston head (Plate 11b) to the threaded end of the piston rod, the thread is treated with an extremely strong anaerobic adhesive to ensure that the assembly torque is maintained and that the piston is held firmly in position. The larger threaded component (left centre) is the ram cylinder end cap (see Plate 11b). Material: Permabond APP460. (Photograph courtesy JCB Ltd.)

Plate 10b   After the fitting of the gland bearing and piston head (Plate 11a), the assembly is offered to the hydraulic cylinder of the ram body. Once inserted and clear of the internally threaded end of the cylinder, a specifically selected low-strength anaerobic composition is used to seal and lock the end cap (Plate 11a) in place. While quite capable of sealing the threaded joint (135 bar) the material chosen will not prevent the end cap from being unscrewed readily – despite the diameter of the thread – should this be required. Material: Permabond A137. (Photograph courtesy JCB Ltd.)

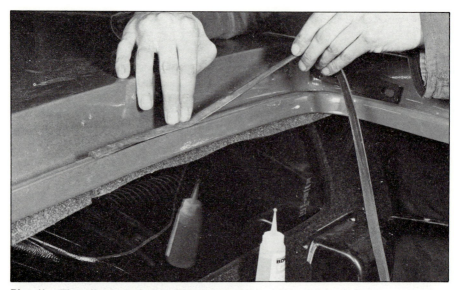

Plate 11   The tail-gate seal of the Esprit Turbo is bonded in position using a cyanoacrylate adhesive. The painted surface of the GRP body is lightly abraded prior to assembly but the rubber itself is not treated in any way. The rubber seal – which is a 'D' section – is stuck down in convenient lengths of about 0.5 m. Material: Permabond C2. (Photograph by Focalpoint, courtesy Lotus Cars Ltd.)

Plate 12   The doors of this Foden S10 Fleetmaster are constructed from bonded aluminium panels. Material: Permabond F241. (Photograph courtesy Fodens Ltd.)

Plate 13   The rubber teething ring is bonded into the PVC air bellows with a cyanoacrylate adhesive. Material: Permabond C2.

Plate 14   The hammer head and shaft are bonded using a rubber toughened, two-part, cold-curing epoxide. This form of assembly prevents the loosening so often associated with conventional wedging techniques. Material: Permabond E32.

Plate 15   The magnetic ferrites of this electric motor are bonded to the external case with a rubber toughened acrylic adhesive. This assembly technique avoids the problems of dimensions and distorted magnetic flux incurred when other methods are used. Material: Permabond F241.

Plate 16   The brass barrel and aluminium outer of this door handle are bonded using a rubber toughened single-part, heat-cured epoxide. This particular grade will flow just like solder prior to solidifying as it cures. This is in direct contrast to the material used in Plate 19 which is formulated so as to maintain a very high viscosity – specifically to prevent unwanted flow. Material: Permabond ESP108.

Plate 17   Power for the Chieftain tank is supplied by the Leyland Vehicles L60 engine whose wet cylinder liners are sealed with Permabond A150. This anaerobic material was specially developed to cope with the extremely hostile environment found within the engine. Extensive proving trials have shown that A150 will withstand the high temperatures and coolant pressures generated during use. (Caption and photographs courtesy of the MoD.)

Plate 18   The window panels of this coach are made from cold-press moulded GRP. They are stiffened with both aluminium and steel formers which are bonded directly onto the panel with a toughened acrylic adhesive. Material: Permabond F241. (Photograph courtesy British Rail Engineering Ltd.)

Plate 19   The steel bonnet of the tractor and its various stiffeners are bonded with a rubber toughened single-part heat-cured epoxide. No special preparation of the surface is required. This particular adhesive will not flow and remains in place as the temperature is raised to induce cure. Material: Permabond ESP105. (Photograph courtesy David Brown Tractors Ltd.)

Plate 20   A general view of a 'Grecon' laminating machine which is used to bond resin-impregnated paper foils to chipboard prior to its subsequent use in furniture manufacture. An emulsion adhesive is applied to the board and dried with infra-red heaters. The foil to be bonded is pressed onto the board with heated rollers. The heat softens the dried adhesive and causes it to form a strong water-resistant bond on cooling. Production rates are high: nine 12′ × 6′ (360 × 180 cm) boards are produced every minute. Material: 083-1217 one-component emulsion adhesive manufactured by National Adhesives & Resins Ltd.

Plate 21   Heat- and water-resistant wood-veneered chipboard panels are produced from board which has been coated on both sides by adhesive applied by this machine. After being coated the board passes on knife-edge rollers to the next station where it is hot-pressed at 100°C to cure the adhesive. The knife-edge rollers are continuously washed in their water bath in order to ensure that they are kept clean. Material: PLYLOCK 3500 two-component emulsion adhesive manufactured by National Adhesives & Resins Ltd.

Plate 22   A close-up of
the complex of feed and
press rolls.

Plate 23   Insulated roof
panels are prepared from
the shaped steel profile
shown and polystyrene
board. Both are sprayed
with a solvent-borne
polychloroprene-based
contact adhesive. By
balancing the rate of solvent
evaporation carefully it is
possible to apply the
adhesive without damaging
the polystyrene – which
dissolves readily in many
organic solvents, Some 5-10
minutes after being sprayed
the two parts are assembled
and passed through rollers
which press them firmly
together. This process is
very quick and a stationary
press is not required.
Material: SPRAYLOCK
901 manufactured by
National Adhesives &
Resins Ltd.

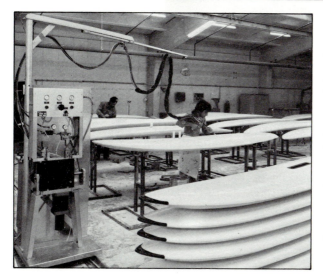

Plate 24   This large
metering, mixing and
dispensing machine will
handle a wide range of two-
component systems –
epoxy, polyurethane,
polyesters and silicone
rubbers. If required it will
deliver pre-set shot sizes.
This particular VARI-
O-MIX 2, Type S is being
used in the assembly of a
centreboard casing during
the manufacture of a
windsurfer. (Photograph
courtesy Prodef Engineers
Ltd.)

3   The special role of anaerobic adhesives in mechanical engineering

3.1 Introduction

Anaerobic adhesives were created specifically for assembling a wide variety of mechanical engineering components. The impact their properties and use have on fundamental design concepts is discussed in relation to other types of adhesives in Chapter 2. What this approach does not demonstrate is the overall value of these unique materials to the mechanical engineer in a wide range of situations, nor does it show the degree to which the characteristics of individual grades are matched to the intended end use.

Anaerobic adhesives come in a wide variety of strength and viscosity combinations. The most popular grades are listed in Table 3.1 which shows that the different viscosities enable the adhesive to cope well with the wide variety of gaps likely to be encountered. Similarly, the various mechanical strength levels available allow for maintenance and dismantling of parts where necessary, or for the permanent assembly of components that will never need to be separated. The successful introduction of toughening into the stronger grades has greatly extended their practical applications in difficult and demanding environmental conditions.

3.2 Applications

The benefits of using anaerobic adhesives are now generally accepted and they are enjoying increasing use. Although individual applications are legion, they fall readily into two major areas. By far the larger is the permanent and semi-permanent bonding of one component within another – co-axial bonding. The other smaller but nonetheless important and growing in volume, is their use as gasketting media.

As will be readily appreciated, the vast majority of fitted mechanical components are co-axial, so it is convenient to subdivide this group into smooth and splined co-axial components, and then further into threaded parts or fasteners.

Table 3.1 Characterisation* of the primary anaerobic adhesives

	Strength		Viscosity mm²s⁻¹ (cSt)			
DTD Number	Nuts & bolts (Nm)	Collars & pins (MPa) (MNm⁻²)	<50 V. low	125±50 Low	500±250 Medium	Thixotropic based on 8 Pa s
5629	2±1 V. low	‡	‡	Grade 1	Grade 2 D	‡
5630	5±2 Low	6±2	‡	Grade 1 B	Grade 2 E	Grade 3 H
5631	10±3 Medium	12±5	‡	Grade 1 C	Grade 2 & Grade 3 § F	Grade 4 I
5632	16±3	21±4	Grade 1†	‡	Grade 2	Grade 3
5633	Min. 19 V. high	Min. 17	Grade 1† A	Grade 2	Grade 3 G	Grade 4

* Crown Copyright.
† Collar and Pin Shear Strength not specified for these grades.
‡ These characteristics have not yet been allocated.
§ Thixotropic version '5631' Grade 2.

Notes:
1 The grades chosen, A-I inclusive, are discussed in 'case histories'.
2 The relevant test methods are given in DTD 5628.

3.2.1 Co-axial – smooth

Perhaps the best example of semi-permanent assembly is bonded bearings. Maintenance may be required, and so adhesive performance must be strictly controlled. If too strong, then impossibly large forces are required for dismounting; if too weak, dynamic forces will cause joint failure in service. Typically, an adhesive for this purpose should generate a resistance to axial static loads of between 10 and 14 MPa. Although impressive, this performance is not the most revealing demonstration of the materials' capabilities. Anaerobic adhesives have been used successfully for many years for the semi-permanent retention of cylinder liners in engine blocks. The load and service conditions which have to be met here can be readily appreciated.

3.2.2 Co-axial – splined

Anaerobics have been found to extend the working life of splined joints – including keys and key ways – since they ensure that assembly does not cause distortion and that the imposed loads are evenly distributed. Typical applications are drive shafts on light cars and heavy vehicles alike and the securing of pump gears.

The use of anaerobics on such components avoids:
Distortion caused by interference, shrink and taper fits;
Stress due to distortion;

Fatigue aggravated by stress;

Fretting caused by chattering, since the adhesive completely fills the gap between the mating parts;

Close tolerances such as those used with interference fits. The ideal gap for the adhesive is 0.05 mm but gaps up to 0.25 mm may be accommodated;

Sleeving, reboring and shimming during maintenance operations.

It also allows:

Lighter sections to be specified by overcoming the problems of force fits, etc;

Reduction in machining time as tolerances are not so tight and surface finish not so important;

Reduced assembly time because of sliding fits;

Less scrap from reduced machining and wider tolerances.

When circumferential shear forces are not too high, an adhesive may replace conventional retention techniques. However, where shear forces are too great for a 'press-fit' to carry, the adhesive should be used with keys or splines. This combined technique still realises may of the above benefits.

3.2.3 Threaded parts

Anaerobic adhesives offer an unrivalled means of locking and sealing threaded parts. Not only are they more economical than conventional retaining and corrosion-prevention methods; but they are particularly useful for locking and sealing pipe joints and securing studs.

In the former, the risk of getting shreds of fibrous packing or tape into the pipe is completely eliminated as is the problem of sealing and setting an elbow joint in a specific position.

The benefits of using an adhesive to retain studs are very important. With interference fits avoided, castings can be thinner; blind holes are unnecessary as the adhesive can seal the thread. However, adhesive selection needs care to ensure that the requirements of any particular application are met.

In some critical assemblies, such as the fitting of an internal combustion engine cylinder head, correct clamping load is vital. Either the traditional torque/tension method, or the more recently developed techniques of differential torque analysis must therefore be used. With torque/tension, the relationship must not be modified by the anaerobic materials used to lock the stud and nut. It has been shown that the relationship remains effectively unaltered if the fasteners are not degreased before assembly. For maximum precision, cleaned fasteners may be used in conjunction with special 'lubricating' anaerobic versions.

3.2.4 Gaskets

Appropriate forms of anaerobic composition make very successful gasket replacements – particularly because they are virtually incompressible. Further-more, the same material may be used to seal several different assemblies unlike the conventional gasketting techniques which are very application specific.

Damage to the seal by excessive component movement under high pressure must be avoided by suitable design and correct selection.

3.2.5 Sealing porous castings and welds

Low viscosity materials offer a quick and simple way of sealing porous castings and hairline cracks. Anaerobic cure follows penetration by capillary action. Other porosity treatment techniques often involve long and complicated curing cycles.

3.3 Anaerobic adhesives in use

The use of anaerobic adhesives should be considered in detail during the design stage (see Chapter 2). Sometimes, however, where this is not done, adhesives are used to remedy bad design or to solve an unexpected production problem. Adhesives may then not give the performance of which they are capable. However, in practice, adhesives' great superiority over traditional methods, such as interference fits, gives them reserves adequate to meet most contingencies.

If adhesives are expected to be used, then two basic issues need to be resolved at the design stage – adhesive strength and viscosity. With these characteristics determined, the most appropriate grades may be chosen readily.

The next point is to decide whether any problems are likely in applying the adhesive; special techniques may be necessary or an applicator desirable. Completely automatic, self-regulating dispensers are available to treat the appropriate area of the components with a precise quantity of adhesive.

One decision still to be faced is to devise a test to indicate satisfactory performance over the intended life of the assembly. Fortunately, sufficient evidence has now been accumulated to enable manufacturers to give positive recommendations in most cases. Occasionally, on fringe applications, it is necessary to advocate long term field trials (see Chapter 6).

Introducing the adhesive onto the production line requires great care as an inadequate understanding of the material and its properties may cause unnecessary difficulties. Common mistakes are the use of too little material and not allowing enough time for cure. Sometimes, quite inappropriate tests are devised and in consequence the performance is misjudged.

These problems can be avoided by working closely with the adhesive manufacturer.

Despite such teething problems once normal production is under way, there are remarkably few problems. Occasionally a change in personnel or in the nature of the components themselves may give rise to difficulties. Labour problems are usually self-evident; problems with materials are often more difficult to identify.

Sometimes, if adhesive cure time is reduced, production continues satisfactorily, followed later by an apparently inexplicable failure. Investigation usually reveals a second and often minor change, perhaps in cleaning methods, that has modified component surface chemistry so that the curing period should be increased rather than cut.

Similar variations stem from temperature changes. Large castings take a long time to warm up after exterior winter storage and accelerators may be needed to compensate for the reduced reaction rate.

Despite these and other similar problems it is clear, from experience, that adhesives can cope very adequately with most industrial conditions and that they usually perform year in and year out with very little trouble.

3.4 Case histories

A few illustrative case histories provide a feel for the application scope of anaerobics in mechanical engineering. The eight examples below cover the main outlets for materials of this type.

Case 1 Mounting electrical rotors ('G' in Table 3.1)

The traditional way of assembling the laminated rotor of an electrical motor onto its shaft is to use an interference fit. Satisfactory results hinge on close tolerances and great care in ensuring that the shaft is not distorted during assembly. The use of an adhesive, especially with an accurate jig, reduces the capital equipment needed and ensures that only stress-free, undistorted assemblies are produced.

Generally a strong, medium-viscosity product will be perfectly satisfactory. However, if liquid adhesive could be sucked into the gaps between laminations, a more viscous or thixotropic grade and possibly an accelerator will prevent loss of adhesive into the laminations prior to hardening.

Case 2 Carburettor sealing ('A' and 'G' in Table 3.1)

A conventional carburettor contains several pipes and tubes pressed into the carburettor body and retained by interference fits. Usually, these tubes carry no significant force and, until recently, the method of fitting was never questioned. However, emission control legislation now covers the loss of petroleum vapour as well as exhaust emission from vehicle engines and accessories – including the carburettor. Enough petrol vapour can escape through an interference-fitted joint to require its sealing. Strong, medium-viscosity materials are particularly effective for this purpose and may be used for all but the most extreme conditions. Where temperatures are liable to exceed 150°C for prolonged periods, the special heat-resisting grades should be used.

Case 3 Mounting bearings and bushes ('F' and 'G' in Table 3.1)

As in Case 1, interference fits are usually used to mount bearings on a shaft and to retain both bushes and bearings in a housing. This implies a level of capital expenditure on grinding equipment and presses, yet does not avoid component stressing and distortion. In contrast, use of an adhesive allows stress-free, undistorted assembly to be carried out without expensive equipment. Adhesives of medium strength and viscosity have been formulated especially for applications of this type. The adhesion developed after hardening is similar to the frictional resistance achieved by interference-fitting the same components. Hence, satisfactory in-service performance can be confidently anticipated and conventional equipment can still be used to separate the components for maintenance. A particular advantage of adhesives is that they allow the use of much lighter housings, which can be designed to withstand service loads alone, rather than to be strong enough to hold the bush or bearing in position. This contrasts directly with press-fitted assemblies where extra housing thickness is essential to prevent bearings from moving out of position. In such situations, a high-performance adhesive should be used so that its bonding strength will compensate for the lower elasticity of an aluminium housing.

Case 4 Permanent locking of threaded fasteners ('G' in Table 3.1)

The mounting of a trolley castor is a perfect example of the permanent locking of a threaded fastener.

The problem here was that the nut retaining the castor axle in position and the screw holding the castor onto the main trolley frame were frequently loosened by vibration when the trolley was used on rough surfaces. A strong adhesive completely overcame the problem, though a weaker grade would almost certainly have prevented loosening. However, the stronger grade was chosen to foil vandals' attempts to remove the trolley wheels.

Case 5 Sealing threaded pipe fittings ('H' and 'I' in Table 3.1)

Anaerobic adhesives are frequently used as sealants in various applications. Of these, perhaps the most useful illustrations are the sealing of pipes, whether unions are intended to be permanent or dismountable for maintenance.

Conventional sealing compounds present severe disadvantages compared with the performance of anaerobic adhesives. Fibrous packing, mastic or tape, almost invariably require the assemblies to be tightened to the limit of the thread to effect a seal. Besides being difficult this often leads to mis-alignment.

Using anaerobic adhesives as sealants makes correct alignment much easier to achieve. Furthermore, they avoid the common problem of solid matter, fibrous packing, mastic or tape entering the pipe – especially important with delicate systems.

Unlike combinations of hemp and mastic, anaerobic adhesives perform consistently in their role as sealant. Furthermore, they do not deteriorate with age as do many traditional products, some of which also support microbiological activity.

Case 6 Locking threaded fasteners where maintenance may be required ('B' 'C' 'D' 'E' and 'F' in Table 3.1)

Many threaded fasteners have to be removed during maintenance. Several grades of adhesive are available to make this possible without mechanical damage.

However, it is not generally realised that such adhesives can be used to lock threaded fasteners and yet allow subsequent adjustment. A typical example is an adjuster screw mechanism of a reed switch. Other applications include tuning circuits in electronic components and adjusters in hydraulic brake mechanisms.

Medium-strength adhesives have now been proven to out-perform all normal mechanical means of retaining threaded components while still allowing simple removal of the fasteners for maintenance.

Case 7 Sealing plane surfaces ('H' and 'I' in Table 3.1)

Traditional methods of sealing surfaces are effective in undemanding conditions but severe environments may cause their failure. This will be accelerated by: solvent loss, the formation of voids, viscous flow, solubility and lack of chemical resistance.

Anaerobic adhesives used as sealants do not suffer from such problems. Very viscous, though thixotropic, grades give immediate sealing for pressures up to 30 bar – even before hardening. After hardening they will cope with very much higher pressures. Lower viscosity grades should not be pressurised before curing but the time taken to harden may be reduced by the use of an accelerator spray or by heating.

Low-strength thixotropic adhesives were used successfully to seal the halves of a gearbox which had to be pressure tested immediately after assembly. The weak grade chosen allowed ready removal of the lid during maintenance and its high viscosity prevented leakage even before completely cured.

Case 8 Sealing porous castings and welds ('A' in Table 3.1)

Because they are metal catalysed and polymerise spontaneously in the absence of oxygen, anaerobic materials are ideally suited to the sealing of porosities. Unlike traditional sealing media, there are no volumetric changes associated with solvent loss, nor is heating needed to harden the adhesive. Low viscosity, high-strength products should be used.

3.5 Application techniques and general comments

3.5.1 Loss of adhesive

Generally, very few problems are associated with the use of anaerobic adhesive and most are usually caused by inadequate application or adhesive loss during, or prior to, assembly.

Atmospheric oxygen within a joint inhibits setting and so the adhesive may not harden quickly unless it fills all the voids. Even with enough liquid applied, some may have been lost before the adhesive hardened or, alternatively, the wrong assembly technique may have displaced the liquid. A study by a major company on the assembly of bearings illustrates the point. Applying the adhesive to the face of the female component was found to give higher and more consistent performance than the alternative technique of applying the adhesive to the shaft, even though this gave quite acceptable performance in normal production.

Blind tapered holes can present difficulty because the air pressure generated by stud insertion tends to force the liquid out. Four techniques overcome this and the most appropriate to the individual application should be chosen.

After liberally applying the adhesive to the hole and stud, the stud should be screwed slowly and gently into the hole as far as possible and then screwed back. The vacuum created will then draw adhesive from the fillet at the entrance to the hole back down the thread ensuring that it is adequately filled.

A narrow slot may be machined along the longitudinal axis of the stud to act as an air vent.

An excess of adhesive should be placed at the bottom of the hole so that the descending stud forces it through the engaged threads.

A similar effect is achieved by placing the adhesive about two thirds of the way down the hole.

3.5.2 Difficult materials

Even with the correct application technique and complete void filling, observed performance may be lower than that suggested in the standard data for a number of reasons:

The surfaces are dirty and should be cleaned;

The surface may be exceptionally smooth;

The surface may be easy to strip away, eg zinc or cadmium plating;

The 'engineering' thermoplastics used are often difficult to bond;

Soft metal is used for one or more of the components.

There is an increasing need to bond small plastics parts – fans, pulleys, light gears and the like – onto a high quality, finely finished, steel shaft. Because the smooth surface offers little opportunity for mechanical engagement and most of the 'engineering' thermoplastics are difficult to bond, much reduced performance is common.

If it is necessary to maximise performance, the engaged area should be as large as allowed by the design and the plastic's surface should be chemically treated or roughened if at all possible. Finally, a high-performance adhesive of low modulus – preferably toughened – should be used. The selection procedure of Section 5.2 helps locate a suitable material. Note: in the presence of a steel part, most high quality anaerobic adhesives will set satisfactorily – even if one surface is non-metallic – but two non-metallic surfaces will invariably require a primer if an anaerobic adhesive is to cure.

From time to time extremely difficult metallic surfaces inhibit the curing reaction which can normally be restored with an accelerator.

3.5.3 Solvents and solvation

Liquid anaerobic adhesives dissolve in most organic solvents and oils and, even when true solution is not possible, they normally disperse harmlessly. With certain exceptions, the liquid adhesives will dissolve or swell many thermoplastic materials such as: Polyacrylates, Polyvinylchloride, Polystyrene and, after lengthy exposure, natural rubber and shellac.

In fact, all solvent-sensitive or soluble material will be either dissolved or swollen on exposure to the liquid adhesive; the hard adhesive is, however, inert. Some thermoplastic materials are unaffected by the liquid adhesives. These are: Polyacetal, Polyethylene, Polypropylene, Nylon (polyamide) and Teflon (PTFE).

Thermoset plastics are undamaged by the liquid adhesive, and these include: Glass fibre polyester composite, Epoxies, Polyurethanes, and Bakelite materials based on phenolic resins.

3.5.4 Accelerated hardening

All chemical reactions are speeded by increased temperature and every increase of 8°C doubles the cure rate of an anaerobic adhesive. Curing can also be speeded by the presence of catalysts, often available as aerosol sprays.

Such accelerators are useful where production conditions severely limit the time available for the adhesives to harden, but their use should not be abused. Excessive application can cause reduced adhesive strength.

3.5.5 Effect of heat

Increasing the temperature of components assembled with an anaerobic adhesive has three different and quite distinct effects:

> Liquid adhesive hardening rate increases rapidly.
> Hardened adhesive begins to soften in the region of 90–100°C.
> Chemical degradation takes place above 200°C.

Because a hardened adhesive begins to soften around 90–100°C, standard types should only be used as sealants above 150°C and 200°C should be regarded as the maximum design temperature. Heat resisting grades, however, display considerable mechanical strength over the whole range of −55°C to +200°C.

The rate of chemical degradation above 200°C as well as general performance depend largely upon the components used, the working environment and the adhesive type employed. Some of the newer, toughened, anaerobic adhesives will seal cylinder liners adjacent to exhaust ports, because, although decomposition takes place, degeneration is not so rapid that the adhesive is completely destroyed before natural deposits have built up to maintain the seal.

Very short excursions of temperature above 200°C will not prevent the performance of most anaerobic compositions from returning to normal on cooling.

3.5.6 Setting time

At room temperature, a good conventional anaerobic adhesive begins to harden within a few minutes of assembly and components can often be handled quite satisfactorily after 10–20 minutes. However, without an accelerator or elevated temperature, full adhesive strength will not develop for one to two hours. Hence normal batch testing is usually carried out after three hours but for full standardisation and type testing a 24-hour cure time is allowed.

Great care is required for both comparative testing and standardisation, where it is vital to avoid subjective assessment, particularly where measured performance depends largely on variables such as the: surface and its finish; ductility of the metal involved; degree of contamination; distance between mating surfaces; temperature; viscosity of the adhesive; and time for curing.

3.6 Performance characteristics

Over the last few years, the Ministry of Defence has reviewed the characteristics of the more commonly used anaerobic materials. Most known applications can be covered by the properties of only 16 grades and a summary of the major characteristics described in MoD DTD specification 5629-33 inclusive is given in Table 3.1. Table 3.2 illustrates a simplified range for less sophisticated applications. Further simplification is illustrated in Table 3.3.

Table 3.2 Typical application of the ten primary grades of the Ministry of Defence anaerobic characterisation

Use	DTD no	Grade no
Screw locking and sealing	5630	2, 3
Locking and sealing nuts and small pipe unions	5631	1, 2, 3, 4
Bearing retention (removable)	5632	2
Stud locking and sealing	5633	3
Permanent high strength retention	5633	3, 4
Penetrating adhesive	5633	1
Pipe sealing (large unions)	5630	3
Gasket replacement and flange sealing	5631	4

Table 3.3   Classification and employment of the five most important grades drawn from Table 3.2

Use	DTD no	Grade no
Locking sealing screws, nuts and small pipes unions and the retention of removable bearings	5631	2
Stud locking and sealing and permanent high strength retention	5633	3
Penetrating adhesive	5633	1
Pipe sealing (large unions)	5630	3
Gasket replacement and flange sealing	5631	4

4 Surface preparation

4.1 Introduction

Quite contrary to popular belief – a belief founded on the use of traditional adhesives – reliable joints can be obtained from unprepared surfaces. Of course, there is no denying that the better the preparation the better the overall performance. But, providing contamination is not gross, perfectly adequate levels of performance can normally be obtained from: Anaerobic; Cyanoacrylate; Plastisol; Toughened acrylic and Toughened, heat-cured, epoxide-based adhesives.

The 'accommodation' of surface contamination shown by some adhesives depends upon two distinct factors. First, as the anaerobic adhesives rely almost exclusively on their jamming action, any further reduction in their levels of true adhesion is unimportant provided that the contamination is not so gross that the hardened film of adhesive slides on the oil film. Situations as bad as this are rare and, when likely, must be either prevented or countered by cleaning.

By contrast, the other adhesives are either excellent solvents, or become so with the aid of added 'scavengers' when the temperature is raised. This is the main reason why heat-cured epoxies cope so much better with contamination than cold-cured variants. At room temperature the solubility of oil in all epoxy resins is low.

While both cyanoacrylate and toughened acrylic adhesives contain no solvents – all the liquid present is converted to solid – their common base material is an excellent solvent in its own right and so both function well on unprepared surfaces. Toughened acrylics are noteworthy in this respect, though the tolerance of the cyanoacrylates is partially limited because contamination may inhibit hardening. Nonetheless, the cyanoacrylates generally cope well with the unprepared surfaces of the small plastics, rubber and metal parts that they are usually used on.

Plastisol adhesives differ because the plasticising oils they contain become very powerful solvents as the curing temperature is approached. During curing these specialised oils and the contamination they pick up are incorporated in the hardening adhesive mass.

However, the subject of surface preparation should be treated with respect, and careful exploration should still be carried out in spite of the fact that preparation may be unnecessary. In general and taking the non-specialised adhesives into account, surface preparation can be graded to give:

Optimum adhesion with good environmental resistance;

Good adhesion with moderate environmental resistance;

Moderate adhesion with low environmental resistance.

Typically, and for most materials, these levels arise from:

Some form of chemical pre-treatment;

Surface abrasion and degreasing;

Degreasing only.

As pointed out, it would be unwise to settle for a potentially poor performance without considering the overall situation very carefully. For example, some materials are notoriously difficult to bond and perform badly almost irrespective of what is done. By contrast, other materials give good reproducible performance with minimal attention to preparation.

The following detailed discussion of the suitability for bonding of common metals and plastics is summarised in Tables 4.1 and 4.2.

Table 4.1   Suitability for bonded assembly – metals

Material	Suitablity rating*	Comment
Aluminium and its alloys	2 – 3	Load-bearing but surface must be properly prepared for use in severe environments
Copper and alloys	3 – 4	Do not load. Careful preparation necessary
Steel (mild)	1 – 2	Load bearing. Really throrough preparation is not always required
Steel (HSLA)	1 – 2	Load-bearing. But experience limited as yet
Steel (stainless)	2 – 3	Load-bearing. Careful surface preparation may not be necessary
Steel (precoated PVC or paint)	3 – 4	Cannot be loaded but can bond well
Zinc or zinc plating	2 – 3	Load with care. Preparation necessary
Zinc passivate	2 – 3	Load with care – correct passivate necessary

*Scale: 1 good, 2 problems, 3 many problems, 4 not recommended.

Table 4.2   Suitability for bonded assembly – plastics

Material	Suitablity rating*	Comment
Plastics		
Fibre re-inforced GRP		
theremoplastic	?	Only load lightly
thermoset	1 – 2	Load-bearing
acrylic faced	2 – 3	Do not load
wood faced	1 – 2	Can be loaded
Pultrusions	1 – 2	Load-bearing
SMC	1 – 2	Load-bearing
CFRP	2	Load-bearing
Other Plastics		
ABS	3 – 4	Only load very lightly
Nylon	2 – 3	Load lightly
Polyolefin } Plus variants on basic type all of which	4	Do not load
} may or may not be filled to some degree		
Polyurethane }	3	Light loads
PVC (rigid)	2 – 3	Very light loads

*Scale: 1 good, 2 problems, 3 many problems, 4 not recommended, ? not enough known.

4.2 Common materials and their suitability for bonded assembly

The selection procedure of Section 5.2 depends largely on adhesive/adherend compatability. Information included there gives a good insight into the interdependence of each on the another. However, the following general review of the more common materials may be helpful.

4.2.1 Aluminium and its alloys

The clean appearance of an aluminium surface is very deceptive; it is actually composed of a thin layer of oxidation products, invisible to the unaided eye, which forms a weak link between the true aluminium surface and the adhesive. Although the oxide films on some alloys is reasonably stable, and gives fairly robust joints with some adhesives without preparation, this must not be generally assumed but should be checked and evaluated. Failure to remove a weak oxide layer will always result in a weak joint.

The oxidation of aluminium in atmospheric oxygen is extremely rapid, commencing immediately a fresh surface is formed by abrasion. Such surfaces must therefore be bonded as soon as practicably possible.

Even the strong, cohesive and firmly bonded deposits left by acid etching degrade and, if assembly cannot take place immediately, the surface must be protected by some form of coating.

4.2.2 Conventional steel alloys

The wide variety of mild steel alloys available may usually be bonded readily.

4.2.3 Zinc-plated steel sheet

Galvanised surfaces have proved notoriously difficult to bond. Wherever possible, zinc should be avoided for it readily forms a wide range of weak surface oxidation products. Even with these removed by abrasion, the zinc layer itself remains a potential source of weakness and a good adhesive will frequently pull it away from the underlying steel. Nonetheless, correctly applied zinc plating can be successfully bonded with the correct adhesive after chemical passivation of the zinc.

4.2.4 Passivated zinc-plated steel sheet

Numerous investigations have shown that variable results are obtained from conventional zinc passivates. While these corrosion-inhibiting surfaces are perfectly adequate for their intended purpose (and usually for painting) the majority are either unsuitable or unreliable for bonding. However, refined passivation techniques give consistent results and, before commitment is made, specialised advice should be sought.

4.2.5 PVC-coated steel sheet

While structural joints cannot be formed on the PVC cladding of mild steel sheets, the toughened acrylic adhesives give good adhesion on the PVC.

4.2.6 Painted steel panels

Pre-coated steel surfaces can be effectively bonded, though not for structural work, by a variety of adhesives. The best results are likely from those which accommodate flexing of the thin sheet.

4.2.7 Stainless steel alloys

Stainless steel alloys have often proved difficult to bond, but the introduction of the toughened adhesives has alleviated the problem. Both the toughened acrylic and single-part epoxy types will bond these alloys well. Abrasion followed by a solvent wipe appears to be sufficient in most cases, though chemical treatments may be beneficial in more demanding environments.

4.2.8 Thermoset GRP

This can be bonded readily with polyurethane adhesives and particularly with toughened acrylics and epoxies. The surface-activated versions of toughened acrylics may not cope too well with the gaps found in larger structures but the recently developed acrylics, which require mixing, cope easily.

4.2.9 Acrylic-faced thermoset GRP

These materials are probably best bonded with acrylic-based toughened adhesives. Both the adhesive and its associated initiator, in their liquid forms, are likely to cause stress cracking in prolonged contact with the acrylic face. These adherends should therefore be bonded using pre-mixed adhesive/initiator if at all practical. Joints based on the acrylic face should never be used for structural assembly work.

4.2.10 Wood-faced thermoset GRP

This structural material can be bonded extremely well with a wide variety of adhesives depending upon the face being bonded.

4.2.11 Pultrusion and SMC – see Thermoset GRP (4.2.8)

4.2.12 CFRP

These structural composites, often epoxy-based may be bonded readily with a range of adhesives – preferably the toughened types.

4.2.13 ABS

ABS is notoriously difficult to bond. Although toughened acrylics can be used, stress cracking is likely if uncured adhesive or free initiator remain on its surface. Pre-mixed versions do not appear to cause this problem.

4.2.14 Nylon

As a general rule, nylon surfaces demand great care in preparation for bonding.

4.2.15 Polyolefin

These materials cannot be bonded satisfactorily without the appropriate preparation technique.

4.2.16 Polyurethane

The increasing use of reaction injection moulded (RIM) and similar poly-urethanes in motor vehicle components has stimulated interest in bonding this polymer. It has not proved particularly easy and so far the best results are from polyurethane adhesives and, to a lesser extent, from toughened acrylics. Often, preparation is required prior to the urethane adhesive being used – regrettably with apparently physiologically active materials.

4.2.17 PVC

Both rigid and flexible PVC may be bonded with toughened acrylics, cyanoacrylates and the solvent-based adhesives. However, although not as convenient as the cyanoacrylates, the toughened acrylics give the best overall performance particularly in severe environments.

4.3 Preparative techniques

While surface preparation is not always required – some adhesives cope extremely well with poor surfaces – simple steps such as cleaning and degreasing are often needed. Components made from materials with weak or loose surface layers, or prone to stress cracking, solvent attack or water migration require special treatment. Techniques to ensure adhesion with even the most difficult of materials are available and widely published; Table 4.3 reviews some general methods suitable for common materials.

Table 4.3 Recommended procedures for preparing the surfaces of materials for adhesive bonding

Material	Cleaning	Abrasion or chemical treatment	Procedure
ABS (acrylonitrile-butadiene-styrene) plastics	Degrease with detergent solution, except for cyanoacrylates – when cleaning and other preparations are probably unnecessary	Etch in a solution of Parts by weight Water — 30 Conc. sulphuric acid (s.g. 1.84) — 10 Potassium dichromate or sodium dichromate — 1 Method: add the acid to 60% of the water, stir in the sodium dichromate and then add the remaining water ADD ACID TO WATER NEVER VICE VERSA	Immerse for up to 15 minutes at room temperature Wash with clean, cold water, followed by clean, hot water Dry with hot air
Aluminium and alloys	Degrease with solvent	Etch in a dichromate solution Prepare as shown for ABS	Heat the solution to 68.3°C ± 2.7°C (155°F ± 5°F) Immerse for 10 minutes Rinse thoroughly in cold, running distilled or de-ionized water Air-dry, oven-dry or use infra-red lamps at not over 66°C, (150°F) for about 10 minutes Treated aluminium should be bonded as soon as possible and should never be exposed to the atmosphere of a plating shop. Even a brief exposure will reduce bonding strength Care should be taken in handling as the surfaces are easily damaged. Bonding surfaces should not be touched (even with gloves) or wiped with cloths or paper
Asbestos board and asbestos cement	Degrease with solvent	Abrade to remove all dust and contaminants HANDLE ASBESTOS WITH GREAT CARE	Repeat degreasing Let the board stand for several minutes so that solvent can evaporate. Then start bonding operation
Beryllium	Degrease with solvent	Etch in a sodium hydroxide solution To prepare: dissolve sodium hydroxide in an equal weight of water, then add enough water to reduce the total concentration of sodium hydroxide to 20% (by weight)	Heat the sodium hydroxide solution to 82°C ± 2.7°C (180°F ± 5°F) Immerse the beryllium for 3 minutes Rinse under cold, running distilled (or de-ionized) water Oven-dry for 10-15 minutes at 150°C to 177°C (300°F to 350°F)

Material			
Bitumenized pipes	Degrease with detergent solution	Abrade with emery paper	Repeat degreasing Rinse well in running water Dry pipes before bonding
Bricks and fired, non-glazed building materials	If contaminated degrease with solvent	Abrade with a wire brush Remove all dust and contaminants	
Cadmium	Degrease with solvent	Abrade with emery paper	Repeat degreasing This metal is normally made bondable by electroplating with silver or nickel, or by passivation
Carbon	Degrease with solvent	Abrade with fine-grit emery paper	Repeat degreasing. Ensure all solvent has evaporated before bonding
Cellulose plastics	Degrease with methyl alcohol, or isopropyl alcohol	Roughen the surface with fine-grit emery paper	Repeat degreasing If using epoxies, heat plastics for 1 hour at 93°C (200°F) and apply adhesive while still warm N.B. follow manufacturers instructions to avoid premature curing of epoxy adhesives
Ceramics – porcelain and glazed china	Degrease in a vapour bath, or dip in solvent	Use emery paper or sand blasting to remove ceramic glaze	Repeat degreasing Let the solvent evaporate completely before applying adhesive
Chromium	Degrease with solvent	Etch in hydrochloric acid solution To prepare: mix equal parts of concentrated hydrochloric acid (s.g. 1.18) and distilled water	Heat the solution to 93°C ± 3°C (200°F ± 6°F) Immerse the chromium in the solution for 1-5 minutes Rinse thoroughly with cold, running distilled (or de-ionized) water
Concrete	If contaminated with oil or grease, scrub the surface with a 2% solution of a non-ionic detergent Wash thoroughly with water	Sandblast about $1/16$ inch from the bonding surfaces and remove all dust and contaminants Where the surface has deteriorated, grind or cut through to good material	Remove dust before applying adhesive Ensure concrete is completely dry
Concrete bituminous	Scrub with a 2% solution of a non-ionic detergent Rinse with water from a high pressure hose until the surface is no longer slippery to the touch	Remove excessively heavy layers of oil and grime by sandblasting, grinding or cutting. Remove all dust and contaminants If limestone, dolomite or other carbonate aggregates are present, etch the surface with acid	

Table 4.3 (Continued)

Material	Cleaning	Abrasion or chemical treatment	Procedure
Copper and copper alloys Includes brass, bronze	Degrease with solvent	Etch with a solution of Parts by weight Distilled water — 197 Conc. nitric acid (s.g. 1.42) — 30 42% (by weight) aqueous ferric chloride solution — 15 or use a 25% (by weight) solution of ammonium persulphate	Immerse for 1-2 minutes at room temperature Rinse the metal in cold, running, distilled (or de-ionized) water Dry immediately with pressurised air at room temperature N.B. Hot air should not be used as it may stain the surface Immerse for 30 seconds at room temperature Wash copiously with cold distilled (or de-ionized) water Dry immediately with pressurised air at room temperature
Diallylphthalate plastics	Degrease with solvent, unless using cyanoacrylates (see notes)	Abrade with medium-grit emery paper	Repeat degreasing
Epoxy plastics	Degrease with solvent	Abrade with medium-grit emery paper	Repeat degreasing
Expanded plastic (foams etc.)	Do not use solvent	Roughen the surface with emery paper	Remove all dust and contaminants
Furane plastics	Degrease with solvent	Abrade with medium-grit emery paper	Repeat degreasing
Glass and quartz (non-optical)	Degrease with solvent	Etch in a solution of Parts by weight Distilled water — 4 Chromium trioxide — 1 Or, use a silane primer in accordance with the manufacturer's instructions	Immerse for 10-15 minutes at a temperature of 23°C ± 1°C (75°F ± 2°F) Rinse thoroughly with distilled water. Dry for 30 minutes at 98°C ± 1°C (209°F ± 2°F) Apply adhesive while glass or quartz is hot
Glass reinforced polyesters (GRP)	Degrease with solvent	Abrade with medium-grit emery paper	Repeat degreasing
Graphite	Degrease with solvent	Abrade with fine-grit emery paper	Repeat degreasing Allow the graphite to stand to ensure complete evaporation of the solvent
Iron (cast iron)	Degrease with solvent	Grit blast or abrade with emery paper	Repeat degreasing
Lead and pewter	Degrease with solvent	Abrade with medium-grit emery paper	Repeat degreasing
Leather	Degrease with solvent	Roughen the surface with sandpaper	Repeat degreasing. Let the solvent evaporate completely before making the bond

Material			
Magnesium and magnesium alloys	Degrease with cold solvent N.B. it is dangerous to put magnesium alloys in a vapour bath Then immerse, for 10 minutes in this solution. Temperature 71°C ± 8°C (160°F ± 15°F) Parts by weight Water 12 Commercial sodium hydroxide 1 Wash with clean, cold running water	Etch in a solution of Parts by weight Water 123.0 Sodium sulphate (anhydrous) 1.8 Commercial calcium nitrate 2.1 Chromium trioxide 24.0	Immerse the solution for 10 mins at room temperature Wash with cold water, followed by distilled (de-ionized) hot water Dry in a hot air stream Apply the adhesive as quickly as possible
Melamine and melamine-faced laminates including Formica, Warite etc	Degrease with solvent	Abrade with medium-grit emery paper	Repeat degreasing
Nickel	Degrease with solvent	Immerse for 5 seconds in a concentrated nitric acid solution (s.g. 1.41) at 25°C (77°F)	Rinse the metal thoroughly in cold running, distilled (or de-ionized water Dry with hot air
Nylon	Degrease with solvent	Roughen the surface with medium-grit emery paper	Repeat degreasing
Paper laminates	Degrease with solvent	Abrade with fine-grit emery paper	Repeat degreasing
Paper (unwaxed)	Do not use solvent	Requires no treatment before bonding	
Phenolic, polyester and polyurethane resins	Degrease with solvent	Abrade with medium-grit emery paper	Repeat degreasing
Plaster	If plaster is fresh, allow to dry thoroughly	Smooth the surface with fine-grit emery paper	Remove all dust before applying adhesives
Platinum	Degrease with solvent		
Polyacetals	Degrease with detergent solution	Etch in a solution of Parts by weight Water 33.0 Conc. sulphuric acid (s.g. 1.84) 184.0 Potassium dichromate or sodium dichromate 1.43 ADD ACID TO WATER NEVER VICE VERSA	Immerse for 5 minutes at room temperature Wash with clean cold water followed by clean hot water Dry with hot air

Table 4.3 (*Continued*)

Material	Cleaning	Abrasion or chemical treatment	Procedure
Polycarbonate, polymethyl-methacrylate (acrylic) and polystyrene	Degrease with methyl alcohol or isopropyl alcohol	Abrade with medium-grit emery paper	Repeat degreasing
Polyester plastics	Degrease with solvent, except when using sensitive materials which require detergent	Roughen with emery cloth or etch in a sodium hydroxide solution (20% by weight) for 2 to 10 minutes at 71°C to 93°C (160°F to 200°F)	After abrasion, repeat degreasing. After etching, wash thoroughly in cold, running distilled (or de-ionized) water
Polyolefins	Degrease with solvent	Etch in a solution of: Parts by weight Water — 20 Concentrated sulphuric acid (s.g. 1.84) — 184 Potassium dichromate or sodium dichromate — 3	Immerse for 15 minutes at room temperature. Wash with clean, cold water, followed by clean hot water. Dry with warm air – not exceeding 60°C (140°F)
PVC (rigid)	Degrease with methyl alcohol or isopropyl alcohol	If possible, use a medium-grit abrasive	Degrease again. Allow to dry
Rubber	Degrease with methyl alcohol, or isopropyl alcohol	Abrade or cut to obtain fresh surface	
Steel and iron alloys	Degrease in a vapour bath	Sandblast or abrade with medium-grit emery paper. Or, use a silane primer in accordance with the manufacturer's instructions	Repeat degreasing. Assemble and bond quickly to avoid surface corrosion
Stainless steel	Degrease with solvent Remove any surface deposits with non-metallic agents (e.g. alumina grit paper) Degrease again. Then vapour degrease for 30 seconds Wash the metal in this detergent solution Parts by weight Distilled water — 138.0 Sodium hydroxide — 1.5 Sodium metasilicate — 3.0 Tetrasodium pyrophosphate — 1.5 Nansa S40/S (Albright & Wilson) — 0.5 Rinse in cold running tap water and then in distilled or de-ionized water Dry in an oven at 93°C ± 2°C (200°F ± 5°F)	Prepare the following solution Parts by weight Distilled water — 3.5 Conc. suphuric acid (s.g. 1.84) — 200.0 Sodium dichromate — 3.5	Heat to between 60°C and 71°C (140°F to 160°F). Immerse for 15 minutes. Wash with clean, cold water, followed by clean hot water. Dry with hot air

Material			
Stonework	If fresh, allow to dry thoroughly	Abrade with a wire brush	Remove all dust
Tin	Degrease with solvent	Abrade with medium-grit emery paper	Repeat degreasing
Titanium	Vapour degrease with solvent Remove any surface deposits with a non-metallic agent (e.g. alumina grit paper) Heat a sodium metasilicate solution (see stainless steel) to between 71°C and 82°C (160°F to 180°F) Immerse the metal for 10 minutes Rinse in cold, running, distilled or de-ionized water	Using equipment made of polyethylene, poly-propylene or tetrafluoroethylene fluorocarbon prepare the following solution Parts by weight Distilled water　250 Sodium fluoride　10 Chromium trioxide　5 Conc. suphuric acid (s.g. 1.84)　50 Dissolve the sodium fluoride and chromium trioxide in water. Slowly add the sulphuric acid, stirring carefully	Immerse the titanium for 5 to 10 minutes, at room temperature Rinse in cold, running, distilled (or de-ionized) water Oven-dry at between 71°C and 82°C (160°F-180°F) for 10-15 minutes
Tungsten	Degrease in a vapour bath	Abrade with medium-grit emery paper	Repeat degreasing
Wood	Wood with a moisture content of over 20% should be kiln-dried before bonding	Remove contaminated material with a sander, plane or axe. Smooth surface with sandpaper	Remove dust by vacuum
Zinc	Vapour degrease with solvent	Etch in a solution of: Parts by weight Distilled water　80 Conc. hydrochloric acid (s.g. 1.18)　20	Solution temperature 23°C (75°F) Immerse the metal for 2 to 4 minutes Rinse thoroughly in cold, running, distilled (or de-ionised) water Place in an oven, at 66°C to 71°C (150°F to 160°F). Dry for 20-30 minutes Apply the adhesive as soon as possible

5 Adhesive selection

There are some 12 major family groupings of adhesives likely to be of value to the engineer. No 'expert' could be expected to know everything about them – let alone their many sub-groups and individual formulations – so the unassisted engineer is unlikely to make the best choice for his application. Although trial and error may eventually furnish a solution it is inefficient, time-consuming and, in the end, may miss the best option. Indeed, some would-be suppliers have a vested interest in steering an engineer away from a better alternative to their own proposal.

The selection procedure in Section 5.2 is arranged to guide the user towards the most suitable adhesives and their manufacturers. For the best and quickest results study the background information given in the relevant sections of this handbook.

The main value of this selection procedure is in mechanical and structural engineering – the requirements of mass production are given particular emphasis. Experience has shown that, in these areas, the procedure will almost always identify a satisfactory adhesive type.

Elsewhere, the proposals will contain a 'sensible' solution and, as a whole, better the likely performance of most experts – since expertise is often narrow and influenced by commercial considerations. Throughout, every effort has been made to ensure that the information is correct – in the widest sense – and that the procedure does not favour the products of any one manufacturer.

Note: The selection procedure is available as a computer program for desk-top computers. This has proved a valuable teaching and training aid as well as a useful design tool. Freedom from manipulating tables gives the user greater insight into the interaction which determines the final choice.

5.1 The major family groups

For the purposes of this review and the process of selection, it is convenient to classify adhesives into the following broad family groups.

Amino	Hot melt
Anaerobic	Phenolic
Cyanoacrylate	Resorcinolic
Emulsion	Phenolic/resorcinolic
Epoxy	Modified phenolic

Plastisol	Toughened adhesives
Polyurethane	Toughened anaerobic
Solvent-borne rubber	Toughened acrylic
Tape	Toughened epoxy

With the possible exception of the Amino family – extensively used in the woodworking industries – all have an important contribution to make somewhere in the fields of structural and mechanical engineering.

The general nature and primary design criteria of each family are discussed below. But the conventional practice of including specific shear strength values has not been followed because published data, often gleaned from totally different sources, tend to emphasise only the shear strength of a joint.

Quite apart from the fact that these values are often obtained under idealised conditions, the practice is very confusing since factors other than shear strength are usually much more significant. At present, however, these are neither well documented nor based on standardised tests or materials.

A much better insight into the actual performance of an adhesive is offered by the descriptions given below and the selection procedure in Section 5.2.1.

5.1.1 Amino

Description

These urea-formaldehyde adhesives come as water-based syrups or powdered solids for mixing with water. Catalysts may be required.

Cure mechanism

Under the influence of heat and/or a catalyst, a polymerisation reaction takes place, water is produced, and the adhesive cures.

Normal application

These adhesives have few conventional engineering applications, being commonly used in the woodworking industry for bonding and lamination of wood and wood-associated materials.

Design criteria

They are available in different viscosities, reaction speeds, etc to suit a multiplicity of uses. However, they may not be used with metal without special primers, they almost invariably require at least one surface to be porous and will not fill large gaps successfully. Because of their high water content, pressure must be applied during curing. They form strong joints with wood – the adhesive's strength is usually greater than that of the latter, but poor environmental resistance renders them unsuitable for exterior use.

Adherend compatibility

These adhesives are compatible with a wide range of wood and wood-associated materials such as the common laminated plastics sheets and are effective on absorbent protein-based and cellulose- based materials, but unsuitable for thermoplastics. Almost invariably one surface must be absorbent.

Surface preparation

None is normally required, except on metal surfaces, though best results come from cleanly cut, well fitting components.

Major benefits

Although some formulations may require complex handling, these adhesives are normally simple and reliable.

Major restraints

Little application outside the woodworking industries. The need for the surfaces to be clamped tightly and occasionally heated is also limiting. They may not be used on alkaline surfaces. Poor gap-filling and the requirement that one surface should be absorbent further limit their use.

Fault finding

The resins on which these adhesives are based are unstable in the liquid state and care must be taken to ensure that the material is in good condition before use. Otherwise, little difficulty is expected unless the quantity of catalyst is incorrect.

Equipment

Ranges from extremely simple, to complex and expensive, according to the nature and quantity of bonded components being produced.

Health and safety

Based on formaldehyde, these resins and the liquid solutions made from powder should be regarded as being physiologically active and likely to cause irritation and possibly dermatitis.

5.1.2 Anaerobic

Description
Anaerobic adhesives, based on the acrylic polyester resins, are produced in viscosities ranging from thin liquids to viscous, thixotropic pastes. Within each viscosity band, individual formulations are available possessing specific strength characteristics. The whole family is unique in being the only one where inter-related strength/viscosity characteristics are provided by manufacturers.

Cure mechanism
For the whole family, 100 per cent liquid-to-solid conversion occurs by radical polymerisation of the vinyl group of the acrylic ester, a reaction catalysed by metal and inhibited by atmospheric oxygen. Typically, therefore, the adhesives cure only when the treated parts are assembled and air is excluded from the mating surfaces. The cure rate on non-metallic surfaces is generally too low for normal commercial use and here a secondary catalyst in the form of a surface primer is beneficial. With closely fitting parts, one metal surface is usually sufficient catalyst and when both surfaces are metallic, polymerisation is rapid at room temperature. Normally, components may be handled between five and twenty minutes after assembly with full strength after at least an hour – possibly very much longer with some materials.

Normal applications
Joint sealing and retention of co-axial components – often both in the same application – are prime examples. Components can be threaded (screws or pipes), or splined or smooth (eg bearings). Special versions of these adhesives (often not truly anaerobic – in that primers may have to be used) give much higher levels of effective adhesion than the normal materials and so may be used in lap joints. Other versions are formulated as gasketting media.

Design criteria
These adhesives come in a wide range of viscosities and strengths to facilitate assembly and to make dismantling possible. Even the weakest of them will normally be capable of retaining a screwed joint in the presence of severe vibration. Although shear strengths of 35 MPa and above are often quoted, these adhesives should not be constantly stressed beyond 10 per cent of this level and their real strength is in carrying very high compression loads. Their popular use in power trains cuts the weight of splined parts and reduces wear. The truly adhesive versions are not particularly good gap fillers, restricting their use to smaller assemblies. The materials characterised by the MoD Specification DTD 5628-33 inclusive meet most requirements of the mechanical engineer. Within the limitation of joint design, they can be expected to meet any normal pressure requirements.

Adherend compatibility

Anaerobics are compatible with all metals; some, such as copper, accelerate curing while others, such as zinc, may reduce the cure rate of some types and may necessitate the use of a secondary catalyst or primer. Performance falls off with softer and more ductile metals and alloys and particularly on plastics parts. Anaerobic compositions attack solvent-sensitive plastics and should not be used on them, nor are they intended for use with rubbery materials.

Surface preparation

Although oil contamination reduces performance, thin oil films are usually tolerated so well that strict surface preparation for most co-axial assemblies is unnecessary. Wiping to remove excess cutting, lubricating or preserving oil is often sufficient, with degreasing only necessary to maximise co-axial perform-ance. Load- bearing lap joints demand the special adhesive types and good surface preparation.

Major benefits

No other adhesive type has the versatility of this family for assembly (with dismantling possible) of co-axially fitting parts. But for lap joints, the ultimate performance of even the best of the truly adhesive anaerobics is readily surpassed by other groups – particularly the toughened acrylics and toughened epoxies. However, both of these may be less convenient to use.

Generally, the anaerobic family is considered very robust and almost all of its members cope with severe environments extremely well. They are simple and easy to use.

Major restraints

Essentially limited as a class to co-axial mechanical assembly, retention and sealing, they also make good general purpose gasketting media. The cure rate depends upon surface activity and may require a supplementary catalyst. The family copes with the gaps of normal engineering practice. As clearances increase, the anaerobics' capacity to cope well falls rapidly. The majority of materials in the family are only suitable for use in lap joints as gasketting media or to seal a gap. Only the special anaerobic materials can be considered to be true adhesives and suitable for use on unsupported lap joints.

Fault finding

In established production lines, characteristic problems are caused by cure rate variation induced by changes in temperature or chemical activity of the surface. A change in gap may well reduce, or increase the handling time. The liquid adhesives deteriorate under poor storage conditions which may lead to premature polymerisation or a reduction in the curing rate (see Section 3.5).

Equipment
Anaerobics are usually dispensed from purpose-made bottles, cartridges, etc.
Metering and dispensing equipment must be specially designed to prevent
premature polymerisation caused by air exclusion.

Health and safety
No major toxicity problems have been associated with this family of materials,
and their general physiological activity can be considered to be very low.
However, some faster curing systems appear to be more aggressive. Products of
the major manufacturers meet the requirements of the National Water Council
and may be used for sealing pipework carrying potable water.

5.1.3 Cyanoacrylates

Description
Cyanoacrylate adhesives are relatively low viscosity fluids based on acrylic
monomers and characterised by extremely fast rates of cure. When placed
between closely fitting surfaces, some will cure to give a strong joint in two to
three seconds.

Cure mechanism
The whole family relies on anionic polymerisation of the vinyl group of the acrylic
ester, giving a 100 per cent liquid-to-solid conversion. The base (alkali) catalyst
normally takes the form of traces of surface moisture – which are, in fact, usually
slightly alkaline in nature. Acids strongly inhibit curing.

Normal applications
The assembly of all kinds of small rubber, plastics and metal parts are typical and
they are often used to bond lengthy rubber strips and mouldings – particularly
weather seals where direct metal contact is not involved. Occasionally, they are
used in the assembly of all-metal mechanical components such as small bearings
for use in benign conditions.

Design criteria
Although nominally very strong, the thermoplastic nature of most cyanoacrylates
coupled with their low softening point and sensitivity to humidity on metallic
surfaces generally restricts their use to very lightly-loaded assemblies. The
majority are brittle and joints should not be flexed or subject to shock loads unless
at least one of the adherends is sufficiently compliant to absorb the deformation
without transferring it to the bond line. The ability of the higher viscosity versions
to fill large gaps is limited by their cure mechanism and they will not cope with
gaps more than about 0.125 mm. Their outstandingly useful characteristic is the
rapid rate of cure – producing bonds in seconds.

Adherend compatibility
The adhesives will bond almost all materials (though a primer may be needed with some), except polyolefin plastics (eg Polythene) and other low surface-energy types such as fluoropolymers (eg Teflon) and silicone-based rubbers. Alkaline glass may cause premature bond failure and all glasses should be silane primed if at all possible, as this considerably improves the joint's humidity resistance. May stress crack stressed mouldings or susceptible plastics – polycarbonate, for example.

Surface preparation
Generally, moulded plastics parts need no surface preparation. Metal should be de-greased though, again, small components may generally be bonded as delivered. Some rubbers, containing various compounding oils (EPDM) will only bond well on freshly cut or abraded surfaces. Chlorinated cleaning solvents should be avoided as these sometimes leave acid traces which prevent cure.

Major benefits
Speed of cure and almost unrivalled ability to bond many plastics and rubbers. Very clean and easy to use.

Major restraints
The rapid cure rate hinders bonding of large areas (though the bonding of long strips presents no practical problems). They cannot cope with large gaps. These adhesives are liable to brittle failure on stiff adherends and are moisture sensitive when used directly on metal exposed to warm humid conditions.

Fault finding
On established production lines, problems usually stem from changes in dimensions or humidity. Very humid or very dry conditions may cause weak bonds or can prevent curing altogether, as can an inadvertent increase in joint gap.

Oil can lead to poor joints and it should be noted that oil may migrate from compounded rubbers. Acid residues from other processes or from chlorinated cleaning solvents will prevent cure. Performance is affected by poor storage conditions, resulting in either premature polymerisation or a marked loss of cure rate.

Equipment
Usually purpose-made bottles. Metering and dispensing equipment must be designed to prevent premature polymerisation.

Health and safety
All the adhesives in the family bond skin very readily and rapidly, presenting a real hazard which must be anticipated but not exaggerated. Industrial experience shows

that, with correct training and, where necessary, appropriate equipment, the difficulties and danger are minimal. However, several of the more popular materials have irritant vapours which must be extracted under constant use. According to grade, physiological activity varies from low to very low and the family as a whole is not considered toxic.

5.1.4 Emulsion

Description
Emulsion adhesives are generally modest viscosity, milky white dispersions often – though not always – based on polyvinyl acetate with up to approximately 65 per cent solids content.

Cure mechanism
The water carrier is usually absorbed by at least one of the bonded surfaces though evaporation can also play some part. As water is removed, the dispersed plastic phase is forced into contact and congeals; the resulting film binds the surfaces together. Cure rate depends on the speed of water extraction – ultimately, the absorbed water is lost by indirect evaporation. Direct evaporation of the emulsion to form a film is used in bonding of open weave materials but, even so, there is still a slower, subsequent release of absorbed water from the substrates. Normally, the cured adhesive film is completely thermoplastic but special acid catalysts, available for some formulations, will promote cross-links in the film to enhance environmental stability.

Normal applications
Extensively used in the woodworking and packaging industries, the adhesives have applications with fabrics and fabric-based components used in 'engineered' products, for example loud-speaker cones. They also bond ceramics well. Normally used for relatively lightly-loaded materials – preferably in good environments.

Design criteria
The water carrier of the emulsion must be lost either by evaporation or absorption and subsequent evaporation. Generally, therefore, at least one surface must be absorbent. The adhesive may corrode metal. Clamps must normally be used to maintain pressure during absorption. The family as a whole is suitable only for lightly loaded components and, except for the cross-linked (two-part) variants, they should not be exposed to warm wet conditions. Heat accelerates their rate of degradation. Note: paste-like versions may have sufficient tack to hold very light parts together and are therefore not limited to minimal film thicknesses.

Adherend compatibility
Emulsions bond most materials which absorb water. Non-absorbent adherends may be bonded if the other surface provides an escape route for water. Thus, some adhesion can be obtained on most metal and plastics surfaces, though care should be taken to ensure that metals are adequately painted to prevent corrosion. Without special preparation, polyolefin plastics, other low surface-energy types such as PTFE and the silicone rubbers cannot be bonded.

Surface preparation
Surfaces must be free of oil and grease.

Major benefits
These adhesives are particularly suited to large scale, low-cost assembly of wood, fabrics and other absorbent materials. They are very clean and simple to use, effective and present no major problems or hazards.

Major restraints
One surface at least must normally be absorbent, most types are readily damaged by warm, moist conditions and they may not be heavily loaded. Except for the tacky versions, handling times may be prolonged. May corrode metal.

Fault finding
There are no characteristic faults.

Equipment
These adhesives may be applied with various metering and dispensing equipment dependent upon the scale of operation. Small quantities are very conveniently dispensed from purpose- made containers.

Health and safety
There are no particular problems with the single-part members of this family, though the acid catalyst of the two-part version can be dangerous and must be assumed to be physiologically active, corrosive and toxic.

5.1.5 Epoxy

Description
Epoxy adhesives are thermosetting resins which solidify by polymerisations and, once set, will soften but not melt on heating. Two-part resin/hardener systems solidify on mixing (sometimes accelerated by heat), while one-part materials

require heat to initiate the reaction of a latent catalyst. Their properties vary with the type of curing agent – amine, amide, acid anhydride – and the resin type used. Epoxies generally have high cohesive strength, are resistant to oils and solvents and exhibit little shrinkage during curing.

Epoxies provide strong joints and their excellent low creep properties make them ideal for structural applications, but unmodified epoxies have only moderate peel and low impact strengths. Although these properties can be improved by modifying the resin, to produce flexibilised materials with improved resistance to brittle fracture, real toughness is only obtained in the so called toughened adhesives (see Section 5.1.12.3) in which resin and rubber combine to form a finely dispersed, two-phase solid during curing. The epoxy adhesives are an important family widely used in structural work, though their high viscosity tends to limit their use on smaller components.

Cure mechanism
Polymerisation is by either of two processes. In two-component formulations, an addition reaction takes place between the active oxygen of the epoxide group and a labile hydrogen, usually from an hydroxy, amine, amide or acid group in the hardener. These reactions usually proceed at room temperature and may be speeded by warming. Some versions – anhydride cured – need heating. Heat-cured single-component variants polymerise through the intermolecular reaction of the epoxide groups driven by an ionic reaction initiated by any one of several types of catalyst, such as dicyandiamide.

Normal applications
Strength, durability and great versatility give epoxies a wide diversity of applications.

Design criteria
Because of this family's versatility, it is probably more useful to summarise its limitations. They can be brittle, cure rate varies enormously with formulation and their viscosity can make use difficult on very small assemblies. In general, their naturally very high strength may not be modified (reduced) readily. This, coupled with their viscosity problem, has prevented their use in the assembly of mechanisms – especially when dismantling is required – and explains the ubiquitous use of anaerobic adhesives here.

Adherend compatibility
This is excellent, except with thermoplastics and rubber where performance is substantially reduced. Low surface-energy materials (polyolefines, fluoropolymers and silicone rubbers, for example) may not be bonded without special surface preparation.

Surface preparation
This family generally needs good surface preparation, except for a few special formulations and the heat-cured, toughened variants, which have proved very tolerant of contamination. Protein-based preserving oils on the surface, however, pose difficulties.

Major benefits
Strength, durability and general versatility. Unlike the surface-activated adhesives, cure rate is generally quite independent of surface chemistry.

Major restraints
Cure rates are generally low unless the adhesives are heated, and the special formulations which cure quickly at ambient temperatures tend to be rather brittle. The untoughened formulations may have a low peel strength and may exhibit brittle failure on impact. Weighing and mixing of the two-component systems must be accurate and thorough and they can be messy and difficult to clear up after use. The cure rate may vary uncontrollably if large quantities are mixed or if there are large fluctuations in ambient temperature.

Fault finding
Problems usually arise from either poor proportioning, mixing or temperature variations. Adhesive failure is almost invariably associated with contamination.

Equipment
Small quantities may be readily hand mixed and the single-part, heat-cured versions may be applied directly from a tube or cartridge. Metering and mixing equipment for the two-component systems is often complex and expensive.

Health and safety
Contrary to popular belief, epoxy resins are thought to be relatively hazard free, unless swallowed, when they must be considered potentially toxic. Dermatitis and similar problems which occasionally arise with these products, are almost always linked with either, or both, nitrogen-based hardeners or powerful solvents used to clean the mixing equipment. Some amine-based hardeners fume and must be used with ventilation. In the film-based adhesives, the resin and hardener in the thermoplastic film are already partially reacted – this is completed on heating. Materials of this type, where the hardener is effectively 'locked-up', have a much reduced level of physiological activity as also have the single-part, heat-cured adhesives whose polymerisation is initiated by catalyst. This latter group must rank amongst the safest of the epoxies and, generally, of all the high-performance adhesives.

5.1.6 Hot melt

Description
Many synthetic thermoplastic polymers can be used as hot melt adhesives. Such adhesives, which often comprise blends of several materials, are to be found in a variety of forms – tapes, films, powders, rods, pellets, blocks and also liquid, solvent-based versions in which the solvent evaporates after placement but before bonding.

Cure mechanism
The thermoplastic mix of materials in the adhesive melts on heating (usually in the range 65°C–180°C) and solidifies on cooling to form strong bonds.

Normal application
The extremely rapid assembly of all types of lightly stressed components unlikely to be exposed to elevated temperatures or extreme environments.

Design criteria
Component design must accommodate the extremely rapid hardening of the adhesives as they cool. Jigs may be required for accuracy and good access is vital. Gap filling is excellent (especially with the foaming versions). Substrate materials must be chosen to match the adhesive's application temperature. Loads may only be light – especially if exposure to even modestly elevated temperatures is anticipated, or the foamed versions are used.

Adherend compatibility
These adhesives are very versatile and appropriate formulations will even bond the difficult polyolefins. However, they are unsuitable for most rubbers.

Surface preparation
Surfaces must be dry and oil free.

Major benefits
Speed and versatility.

Major restraints
They may harden too quickly for accuracy, cannot sustain high loads and are temperature sensitive. The generally high melt-viscosities may make close work on small objects difficult.

Fault finding
The adhesive may chill too quickly or simply fail to 'wet' a contaminated surface.

Equipment
May not always be required where manual placing of sheet film is sufficient. Otherwise, equipment varies in size and complexity according to application, ranging from modest hand guns to extremely large, complex metering and dispensing equipment coupled to fume and vapour extraction facilities.

Health and safety
The major problem is the possibility of severe burns. According to formulation, the vapour and fumes may vary in physiological activity but, in principle, these must always be extracted.

5.1.7 Phenolic/Resorcinolic

Description
Adhesives of this type are usually medium- or high-viscosity liquids, but powdered and film forms are available. The metal-compatible forms are usually heat cured and, along with the acid-catalysed, cold-curing variants, are often used on wood. Water release during curing demands high pressures to maintain contact between the surfaces being bonded. Those resins based on phenol and resorcinol are very similar in nature but the latter are much more durable and are used in more demanding environments.

A major sub-group, the 'modified phenolics', was developed to reduce brittleness, still a very pronounced family characteristic. This modified, more flexible form, is widely and effectively used in the aircraft construction industry. However, the family as a whole has not proved to be particularly attractive in mass production engineering industries where the curing problems are not easily accommodated. Despite this, their role as laminating and bonding agents in the traditional wood-associated industries and the capability of the modified form to bond metal are important.

Cure mechanism
The adhesives are phenol and/or resorcinol part-reacted with formaldehyde, dissolved in a solvent, usually water; water is also released during the reaction. Up to half of the resulting resin may be solvent – though dry powder and film forms are available. The reaction is completed during bonding and this stage may be initiated either by elevated temperature or a catalyst. More water is released during this process. The cure time can be as short as ten seconds or so, but is usually much longer for most formulations.

Normal application
Bonding and laminating wood, metal and a limited range or thermoplastics. Speciality use with friction materials and ceramics.

Design criteria

The group's fundamental limitations include its cure mechanism and the need to press together the surfaces being bonded to accommodate the volumetric change when water is eliminated either by absorption or by evaporation. Thus, the process lends itself to bonding large flat or gently shaped areas which may be readily compressed and, if required, heated. These adhesives are quite unsuitable for bonding co-axial assemblies.

Unmodified materials are brittle, so when the phenolics' high strength and environmental performance are required and there is any possibility of flexing or impact damage, then the modified versions must be used. With the development of the toughened epoxy and acrylic adhesives (see Section 5.1.12), the phenolics – even the modified versions – should only be used where their excellent durability and heat resistance is vital.

Adherend compatibility

Between them the various sub-groups cope with the common industrial materials, particularly steel and aluminium alloys and with many absorbent materials – especially cellulose-based ones. But with few exceptions, such as the polyamides, the family as a whole cannot cope with most thermoplastics though some of the thermoset plastics (normally in the form of PF – phenol/formaldehyde, UF – urea/formaldehyde, and melamine-based laminates) bond extremely well.

Surface preparation

The surface of cleanly cut wooden components and absorbent materials should need no special preparation, but roughly abraded wood may not absorb the adhesive adequately and performance should be carefully checked if the bonded components are to be used in a demanding situation. Metallic surfaces must be thoroughly prepared by the most appropriate method for the material in question.

Major benefits

High shear strength and excellent durability – especially in harsh environments.

Major restraints

The family is prone to brittle failure and even the much improved, modified versions should be used cautiously if mechanical abuse is likely. The usual requirements of heat-induced polymerisation, high clamping pressures and the need for very careful metal preparation greatly restrict the use of these adhesives in the mass-production, metal-based industries.

Fault finding

Since the adhesives are supplied in a part-reacted state for completion during curing, some types may continue to react during storage – particularly in warm environments – so reducing bonding because the substrate will not be wetted properly. Otherwise, most problems are likely to be associated with improper mixing (where required), poor surface preparation and failures during the heating and pressing stages.

Equipment
Varies in size and complexity according to the formulation and application. Large, complex installations are required for many woodworking applications, but despite this, the use of these materials remains a cost-effective and attractive proposition in this industry.

Health and safety
The various resins, hardeners and fumes encountered must be considered to be potentially irritant, dermatitic and toxic if swallowed.

5.1.8 Plastisol

Description
Usually prepared as viscous, immobile pastes – basically suspensions of PVC particles in a combination of liquids, usually composed of plasticisers and reactive monomers.

Cure mechanism
Heating to around 180°C swells the PVC which absorbs the liquid phase of the suspension. Any monomers present react and cross-link to give a semi-thermoset polymeric mass. The process is irreversible and cooling gives a resilient robust material.

Normal application
Assembly of large metal components, typically the lightly stressed panels of motor vehicles – bonnets, boot lids etc.

Design criteria
Compared with the elegance of use of some adhesives, plastisols are relatively crude and do not lend themselves readily to the assembly of smaller components but enjoy extensive use in vehicle construction for doors, bonnets, boots lids and so forth. They are intended for use with metal and cope readily with the gaps and spaces normally found in vehicle panel assembly. This family of adhesives possesses fairly good shear and peel strengths and are very robust. However, they are not true structural adhesives for their softening point is too low. They are often used in conjunction with spot welding. Difficulties are encountered here because the intense transient heating of the welding chars the PVC and releases acid chlorine gas. Corrosion can occur.

Adherend compatibility
Adhesives in this group are traditionally applied to sheet steel but other adherends can also be bonded successfully. The main limitation is the high curing temperature. The hot, liquid adhesive may dissolve or swell other adherends.

Surface preparation
Not normally necessary.

Major benefits
The whole family is simple and easy to use and their tolerance of poorly prepared surfaces – especially residual press oil – ensures that they offer a robust and reliable performance.

Major restraints
Their thick paste-like nature prevents fine work and their aggressive tendency to dissolve adherends during the heated curing cycle tends to limit their use to fairly large metal components. They may only be used for the assembly of lightly loaded parts and, where true structural loads need to be borne, they are best used to supplement spot welding.

Fault finding
They are not prone to any particular hazards, but prolonged storage may cause settlement.

Equipment
Very simple dispensing apparatus is required but fume extraction must be used during curing. As this is usually carried out in conjunction with paint stoving, extra costs are not normally incurred.

Health and safety
There are no particular hazards associated with the cold, unreacted adhesive though some of the monomers used may be irritant and could be toxic if swallowed. However, the fumes and vapours emitted during the curing process are dangerous and must be removed.

5.1.9 Polyurethane

Description
The family group is named after the polymer type formed on completion of the reaction. The adhesives are usually two-component – one always isocyanate-

based, the other formulated from one of several co-reactants, often amines or glycols. Technically, low viscosities are available but health hazards are associated with the low-molecular-weight reactants, so products are normally only of high molecular weight and, therefore, viscous.

Cure mechanism

On mixing, the isocyanate and hardener undergo a powerful, rapid addition reaction at room temperature, normally resulting in a 100 per cent conversion to the polymeric adhesive form. However, if required, gas-producing reactions can be incorporated to foam the adhesive, often used to both bond and insulate. Single-part versions use atmospheric moisture as the cross-linking agent.

Normal application

The bonding of a wide range of components fabricated from metal, wood, various plastics – both thermoset and thermoplastic – and rubber. The adhesives have general utility but their viscosity and rate of cure can be limiting.

Design criteria

Strong resilient adhesives capable of structural use, though the paste-like constitution of many versions prevents their ready use on small objects. The family has excellent gap-filling characteristics particularly enhanced in foaming forms. However foamed versions tolerate only very light or nominal loading. They should not be used in hot wet environments – particularly on metal surfaces. They have proved particularly useful in the fabrication of composite (GRP) structures.

Adherend compatibility

The family as a whole is very versatile. While performance on some plastics is only modest, the group may be used with advantage, on plastics materials sensitive to stress cracking. The polyolefins and other low surface-energy plastics (such as PTFE and some rubbers) present major difficulties unless specially prepared.

Surface preparation

May be critical and depends to a large degree on the nature of the adherend and the individual adhesive.

Major benefits

The family as a whole will bond a wide range of adherends and the various formulations may be readily manipulated to give excellent gap filling, high reaction speeds, all coupled with a good robust structural performance with enhanced impact and peel resistance.

Major restraints
Both the raw materials and the final polyermic form of the adhesives are sensitive
to moisture-induced damage. Atmospheric moisture reacts readily with the
isocyanate resin rapidy rendering it unusable. The cured adhesive should not be
used for high loads in hot, wet environments. Some of the adhesives may be
difficult to use in small scale work and the larger applications will frequently need
complex mixing and metering equipment. The isocyanate base of the resin can be
dangerous – irritant and toxic. Frequently the adhesives must be used with
primers which are also often highly reactive and physiologically active materials.

Fault finding
The family as a whole suffers from inadequate surface preparation and poor
wetting. Variations in reactivity can be induced by moisture, and temperature
changes will affect critically the timing and efficiency of machine-mixed and
dispensed systems.

Equipment
This may well be complex and costly and the high viscosity of some formulations
makes pumping and dispensing difficult without heated lines. This in turn affects
the reactivity of the system, making timing and mixing critical. Fumes and vapour
must be extracted.

Health and safety
Very high standards of hygiene are required because the isocyanate is
physiologically active. This should be backed up by detailed attention to
ventilation and the alertness of workforce and management.

5.1.10 Solvent-borne rubbers

Description
The family is based on natural and synthetic rubber solutions ranging from
relatively low viscosity solutions to viscous pastes and semi-solids.

Cure mechanism
These thermoplastic and flexible adhesives cure by carrier solvent evaporation
and/or absorption. Some rely on a mobile, tacky film of residual polymer, formed
as solvent is lost. This is capable of migrating into a second such film to give a
coherent structure as the two adhesive films flow into each other. Vulcanised,
heat- cured versions cross-link at elevated temperatures to give very robust
adhesives capable of sustaining substantial loads.

Normal application

Laminating plastics sheet and the bonding of an extremely wide variety of materials – particularly where flexibility is required and the operating environment is not too demanding. Vulcanised versions will perform well in bad environments and will sustain substantial loads.

Design criteria

With the exception of the vulcanised versions, the family should only be considered for very light or nominal loading. The formulations within the family vary widely in tack, durability, strength and resistance to fluids and elevated temperatures. Components must be designed to allow precise placement because instantaneous 'grab' is displayed by many formulations. Adequate means of pressing the parts together after initial contact must often be available. These adhesives tolerate repeated flexing and stretching – the vulcanised versions are frequently used to bond rubber mounting blocks to metal substrates. The thermoset variants are very robust and, used in conjunction with an appropriate surface preparation technique, they will withstand the extremes of environmental stress experienced by road vehicles. They are poor gap fillers and their high viscosity and stringy nature makes them difficult to use in fine work.

Adherend compatibility

Except for vulcanised variants, whose curing temperature reduces their scope, the family as a whole may be used on almost any article not damaged by the solvent used for the rubber. A distinction here must be made between an article itself and the material from which it is made. Used in thick section some plastics will resist solvent damage readily whereas thin sheets are easily damaged.

Surface preparation

Unless the surfaces are badly contaminated, preparation will not be necessary beyond the roughening of fibrous materials such as leather. However, the vulcanised versions may well require special surface preparation, depending on application and operating environment.

Major benefits

The basic forms are effective and very simple to use. The more complex vulcanised versions offer one of the few methods by which resiliant rubber may be bonded securely to metal

Major restraints

In the main, the adhesives are restricted to light or nominal loads in benign environments – vulcanised versions excepted. Solvent loss may present both fire and health hazards. Although the bonding process is almost instantaneous, long periods may be needed for the necessary solvent loss prior to joint closure. This can be inconvenient, especially on large components.

Fault finding
In the simple, evaporating variants, assembly either too soon or too late, is the major hazard. With assembly too soon, the surfaces will not be retained in position becuase residual solvent reduces the adhesive film strength and natural stresses in the adherend may cause joint cleavage. Too much solvent loss prior to closure prevents the adequate flowing together of the coating and adhesion will not occur. Specific surface preparation is essential for the load-bearing, vulcanised formulations.

Equipment
This ranges from very simple to extremely complex depending upon the application and its scale. However, the more complex equipment is usually only used by specialist organisations making components for which these materials were traditionally developed. The mass production of small items may require only simple equipment. If used on any appreciable scale, ventilation must be provided.

Health and safety
The physiological activity of the solvents and the vapours and fumes released during the evaporation and vulcanisation stages of the heat-cured versions must be recognised. Solutions are toxic if swallowed, will cause skin irritation and the vapour will irritate the eyes and induce narcosis if inhaled in quantity. However, used sensibly, these adhesives cause few problems – hence the popularity of the simpler versions for domestic repairs.

5.1.11 Tape

Description
Single- and double-sided self-adhesive tapes come in many forms and sizes, using a variety of backing materials such as – cellulose, polyester, foamed polyurethane and PVC. The adhesive is usually based on a tacky, semi-solid acrylic polymer which may be used with a release strip. Some special tapes are heat activated.

Cure mechanism
Apart from the heated variants, the permanently tacky surface of the tape adheres firmly to most surfaces. However, the common materials can only carry nominal loads, though the heat-induced cross-linked structure of the specialised forms may be lightly loaded.

Normal application
Between them the single- and double-sided and foam-supported versions will

bond almost any object whose surface geometry is not too demanding and where the loads involved are either nominal or very light. Specialised types will cope with greater loads and the more demanding external environments.

Design criteria
The family as a whole presents an effective and fast method of attaching light parts and components where only nominal loads may be sustained. Creep can be a problem at temperatures much above normal atmospheric and solvents must be avoided. Surfaces to be bonded must be clean, flat and curved only in one plane – rough surfaces do not bond well since the tape only attaches at the high spots. While foam-based versions can cope well with large gaps, they will not necessarily hold parts firmly in position because the low-strength foam allows a small degree of movement.

Adherend capability
The family as a whole will bond almost any surface to some degree.

Surface preparation
Surfaces must be clean, dry and dust free.

Major benefits
The main advantages are speed, ease of use and freedom from hazards.

Major restraints
Not even the cross-linked versions can carry substantial sustained loads and the family is particularly susceptible to creep. It is also sensitive to heat and solvents. Tapes are quite unsuitable for use on fitted parts – where one piece fits into another – and are difficult to use on three-dimensional shapes. They are not normally effective on contaminated surfaces.

Fault finding
Failure to adhere is usually associated with dust or oily surface contamination. A common failing in use is caused by plasticising oils migrating from the bonded substrate (usually flexible PVC sheet) into the adhesive layer, softening it and leading to premature failure.

Equipment
Very simple mechanical dispensers are available if required.

Health and safety
The family as a whole is generally considered to be free of user hazard.

5.1.12 Toughened adhesives

Description
Toughened adhesives contain a dispersed, physically separate, though chemically attached, resilient rubbery phase. The toughened concept (see Section 1.2) – in the modern sense – has so far only been successfully applied to two adhesive families – anaerobics and epoxies described in Section 5.1.2 and 5.1.5 respectively. It has also led to the creation of an entirely new species of adhesive – the toughened acrylic – which is discussed in Section 5.1.12.2 below.

5.1.12.1 Toughened anaerobic

The toughened acrylic-based anaerobic adhesives take two disparate forms. The first group being essentially similar to the stronger versions described in Section 5.1.2, though their overall performance is considerably enhanced. The second group, while possessing the anaerobic character, are true adhesives. Toughening gives them enhanced shock and peel resistance so that they may be used to bond all types of lap joint. While they do not rely on a jamming action, they may be used with benefit for the assembly of some mechanical components. These adhesives are more conveniently covered in Section 5.1.12.2 because they are more usually employed in the assembly of small components and structures where lap joints predominate.

5.1.12.2 Toughened acrylic

Description
This rapidly developing group of materials, based on a variety of acrylic monomers, has viscosities which are fairly readily modified giving thin liquids, syrupy resins and thixotropic semi-solids. Some are truly single-component adhesives – the anaerobic versions (see Sections 5.1.2 and 5.1.12.1) – while others require some form of hardener. The hardener may be a surface primer or mixed directly into the adhesive.

Cure mechanism
The curing process depends on the radical polymerisation of the acrylic vinyl group giving total liquid-to-solid conversion. In anaerobic variants, this may be by the same mechanism described in Section 5.1.2 but all respond to a catalytic primer – either a surface initiator or mixed directly into the adhesive.

All these adhesives are intended to cure at room temperature. Some of them should not be heated in the uncured state, while others may be heated or warmed in order to speed the process.

Normal application
Extremely versatile, these adhesives will bond almost any substrate with the exception of rubber-based materials and the difficult thermoplastics such as

polyethylene, etc. They are very robust and will cope with demanding environments.

Design criteria
The properties of the whole group give the designer enormous freedom in the assembly of mechanisms and structures. The various materials possess extremely high strength, impact and peel resistance. They are very durable, though currently the long term performance of some versions at elevated temperatures needs to be monitored carefully – particularly on metals in the presence of water.

The real design problem they pose is the limited number of individual materials so far developed. However, they can meet many demands and should be considered whenever a robust performance is required.

Adherend compatibility
With the exception of the polyolefins and other low surface energy plastics, they cope with almost all common engineering alloys and many plastics. The stress cracking of some plastics is a hazard for some versions of these adhesives. None of them bonds rubbers satisfactorily and polyurethane plastics may prove difficult.

Surface preparation
A feature of these adhesives is their ability to cope with surface contamination, particularly mineral oil. However, for the more demanding applications in severe environments, appropriate preparation is recommended.

Major benefits
The group generally copes well with oily surfaces, is simple to use, robust and some versions are capable of filling large gaps. A variety of cure rates are available.

Major restraints
This relatively new family of adhesives does not yet comprise enough individual formulations to cope with all the potential industrial applications where they could make a major contribution. Gap-filling remains a problem with many formulations and some induce stress cracking in susceptible plastics.

Fault finding
The anaerobic variants are subject to the special problems of that family (see Section 3.5) though, when used with a primer, their function can be readily equated with the non-anaerobic versions. As a group these have proved particularly reliable even on oily and contaminated surfaces though some appear more subject than others to polymerisation inhibition induced by chemical contamination.

Equipment

The equipment required ranges from bottle-based, hand-held dispensers through simple pressure-operated applicators to complex mixing and metering equipment. The latter is used in mass production or where viscous/thixotropic, mixed formulations have to fill large gaps.

Health and safety

Vapour extraction should be used where the more volatile versions are in continuous use or primers are sprayed. To date, their industrial record is good and the materials are not believed to be dangerous. Nonetheless, all the products in the group should be considered to be physiologically active to some degree and appropriate precautions taken.

5.1.12.3 Toughened epoxy

The toughening of epoxy-based adhesives confers a substantial increase in the overall performance of both the two-part, mixed systems and the single-part, heat-cured variants. Peel strength, impact resistance and durability are considerably enhanced without any corresponding fall in shear strength. The difficulty faced by the designer is that the limited number of individual formulations do not yet meet all the requirements which will ultimately come within the general scope of the group. However, whenever the absolute maximum performance is demanded from either a mechanical or structural assembly, the toughened epoxies must be considered because they currently offer the ultimate in adhesive performance. Otherwise, with the exception of the comments above, the observations of Section 5.1.5 apply.

Table 5.1   Production factors affecting adhesive choice

Components
- What is the nature of the surfaces?
- Is surface preparation necessary and, if so, is it compatible with the nature of the components?
- Are the faying surfaces accessible, vertical or horizontal?
- Are the components liable to experience major temperature fluctuations?
- Are major dimensional changes likely to affect materially the separation of the faying surfaces?

Adhesive
- Does it have to be mixed and, if so, is equipment necessary?
- If a mixed system, what is the 'pot' life?
- What is the handling time of the bonded parts?
- Is the adhesive hazardous and if so, is special equipment necessary?
- Will subsequent manufacturing processes conflict with the use of this particular adhesive?

Assembly process
- How much space is required by the components, the assembly process and the curing cycle? Consider storage racks, ovens, fixtures, jigs and all equipment.
- What is the true cost of using a particular adhesive compared with other adhesives or other assembly techniques?

5.2 The selection procedure

Many factors influence adhesive selection and the procedure below will identify the most appropriate family or families for a particular application, but individual formulations should only be chosen in discussions with a manufacturer. Only then can the true interplay between the considerations of Table 5.4 be objectively reviewed along with other significant factors – particularly those involving production.

For example, the specific characteristics and requirements of individual adhesive types must be weighed against cost, space and any conflict of subsequent production processes. Polyurethane adhesives, for instance, frequently need relatively expensive metering equipment and ventilators to remove hazardous fumes. Such costs, coupled with those of jigs, can force a design change or, at least, a reappraisal of other adhesives.

Some subsequent production processes may be detrimental to the adhesive bond already formed. The more important production parameters affecting choice are listed in Table 5.1. (See also Chapter 7.)

Table 5.2   Adhesive codes

Adhesive type	Code	Sub-groups of main types	Code	Special notes (see Table 5.3a)
Amino	a	Cold-cured, two-part	a	26
	b	Heat-cured (or warmed), two-part	b	1, 26
Anaerobic	c	Cold-cured	c	2, 27
	d	Cold-cured plus accelerator	d	2
Cynanoacrylate	e	Cold-cured	e	3, 30
Emulsion/Latex	f	Cold-cured	f	26
Epoxide	g	Single part, liquid/paste form (always heat-cured)	g	4, 9
	h	Single part, tape or film form (always heat-cured)	h	4, 5, 9
	i	Two-part, cold-cured	i	8, 10, 28
	j	Two-part, heated to some degree	j	4, 8, 10
Hot melt	k	Always heat-activated	k	6, 32
Phenolic/resorcinolic	l	Cold-cured, two-part	l	26
	m	Heat-cured, two-part	m	1, 26
Phenolic (modified)	n	Heat-cured, two-part	n	1, 26
Plastisol	o	Always heat-cured	o	2, 4, 32
Polyurethane	p	Cold-cured, two-part	p	7
Solvent-borne rubbers	q	Cold and vulcanised	q	11, 29
Tape	r	Pressure sensitive	r	5
Toughened adhesives		Acrylic-based		
	s	Anaerobic, thermoplastic, cold-cured	s	2, 27
	t	Anaerobic, thermoplastic, cold-cured plus accelerator	t	2
	u	Anaerobic, thermoset, cold-cured	u	2, 27
	v	Anaerobic, thermoset, cold-cured plus accelerator	v	2
	w	Non-anaerobic, two-part, cold-cured – could be mixed	w	2, 11
		Epoxy-based		
	x	Single-part, liquid/paste form (always heat-cured)	x	2, 4,9
	y	Two-part, cold-cured	y	8, 10
	z	Two-part, heated to some degree	z	4, 8, 10

5.2.1 Selecting an adhesive

An elimination questionnaire forms the basis of selecting the best type of adhesive from those discussed in Section 5.1 (summarised in Table 5.2).

Its use will indicate the most suitable, or least objectionable, adhesive for any particular application. The approach cannot cover all eventualities and, as unforeseen problems may arise in certain situations, it is always best to consult the adhesive manufacturer on unusual or doubtful applications.

Table 5.3   Adhesive/adherend compatibility table

Ref no	Material (Adherend)	Reject	Secondary	Primary	Ref no	Special notes
1	Cellulose – board, paper, wood etc	c d o s u	g h x	a b e f i j k l m n p q r t v w y z	1	23
2	Cementitious – concrete, mortar etc., including asbestos sheet	a b c d e h l m n o r s t u v	g x	f i j k p q w y z	2	12, 23
3	Ceramic – ferrite, masonry, pottery	a b c d s u	l m n o	e f g h i j k p q r t v w x y z	3	23
4	Fabric – cloth, felt	a b c d e l s t u v	g h o x	f i j k m n p q r w y z	4	23
5	Friction materials	a b c d o s u	e f h i j k p q r t v w y z	g l m n x	5	12, 23
6	Glass	a b c d f l m n s u	e g h i j k o p q r t v w x y z	–	6	13
7	Leather	a b c d g h o r s u x	e j l m n t v w z	f i k p q y	7	23
8	Metals	a b l	f	all but a b f l	8	14
	Plastics[1]					
9	ABS	a b c d g h j m n o s u x z	f k l q t v	e i p r w y	9	15
	Poly					
10	Acetal	a b c h l m n o s u	d f g k x	e i j p q r t v w y z	10	16
11	Acrylate	a b c g h j l m n o s u x z	d f k p q	e i r t v w y	11	15
12	Alkyd	a b c h l m n o s u	d f g k p t v w x	e i j q r y z	12	16
13	Allyl phthalate	a b c h l m n o s u	d f g k p t v w x	e i j q r y z	13	16
14	Amide	a b c h o s u	d f g i j p t v w y z	e k l m n q r x	14	16
15	Amino	a b c h l m n o s u	d f g k p t v w x	e i j q r y z	15	16
16	Carbonate	a b c d h l m n o q s t u v w	e f g j	i k p r x y z	16	17
17	Epoxy (including fibre re-inforced laminates)	a b c l m n o s u	d f k t v	e g h i j p q r w x y z	17	16
18	Ester (thermoset and re-inforced laminates)	a b c l m n o s u	d f i j k t v	e g h p q r w x y z	18	16
19	Ethylene	a b c h l m n o s u	d e f g k p t v w x	i j q r y z	19	18
20	Imide	a b c l m n o s u	d f t v w	e g h i j k p q r x y z	20	16
21	Methyl methacrylate	a b c g h j l m n o s u x z	d f k p q	e i r t v w y	21	15
22	Phenolic (including laminates)	c s u	a b d f l m n t v	e g h i j k o p q r w x y z	22	16
23	Phenylene Oxide	a b c l m n o s u	d f g h k t v x	e i j p q r w y z	23	20
24	Propylene	a b c h l m n o s u	d e f g k p t v w x	i j q r y z	24	18
25	Styrene (including foam)	a b c g h l m n o s u x	d j k p q t v w z	e f i r y	25	19
26	Sulphone	a b c d g h k l m n o s u x	e f i j p w y z	q r t v	26	20

1 Some plastics eg polyethylene, polypropylene, PTFE and silicone rubbers are particularly difficult to bond – except when used in co-axial (slip fitted or threaded) joints.

Table 5.3   (*continued*)

Ref no	Material (Adherend)	Reject	Secondary	Primary	Ref no	Special notes
27	Tetra-fluoroethylene	a b c h l m n o s u	d e f g k p t v w x	i j q r y z	27	18
28	Vinyl Chloride	a b c d g h l m n o s u x	f i j k p t v y z	e q r w	28	–
29	Urethane (Elastomers including foam)	a b c d g h j l m n o s u x z	e f i k q t v w y	p r	29	18 19 23
	Rubbers					
30	Butyl			e q r	30	21
31	Chloro-sulphonated polyethylene			e q r w	31	21
32	E P M			e q r	32	21
33	E P D M			e k l p q r	33	21
34	Fluorinated and other speciality types			e q r	34	25
35	Chloroprene			e i j n p q r y z	35	21
36	Cyclized			e i j n q r y z	36	21
37	'Hard' Structural	Reject all but primaries indicated	Not applicable	e i j n p q r y z	37	21
38	Natural			e p q r	38	21
39	N B R			e i j n p q r y z	39	21
40	Neoprene			e i j k n p q r y z	40	21
41	Nitrile			e k q r	41	21
42	S B R			e i j n p q r y z	42	21
43	'Soft' non-structural			e i j n p q r y z	43	21
44	Silicone			–	44	20

Table 5.3a   Special notes to Tables 5.2, 5.3 and 5.4

1 Hot presses are expensive. **2** Often used without rigorous surface preparation. **3** Plastics and rubber parts not normally cleaned and never with chlorinated solvent. **4** Inexpensive heating techniques available. **5** May need pre-shaping; curved surfaces difficult. **6** Equipment inexpensive. **7** Expensive equipment necessary. Severe physiological hazard. **8** Small volumes may be hand-mixed. **9** Physiological activity very low. **10** Possible physiological hazards. **11** Vapour extraction often needed. Some materials have hazardous flash points. **12** Resistance to cleavage and tensile forces low due to nature of adherend's surface. **13** Glass joints only reliable in dry environments. Otherwise, surface preparation or specialised adhesives necessary. UV curing adhesives are especially useful and will also bond some transparent plastics. **14** Most metal components will bond well in co-axial (slip-fitted or threaded) joint configurations – usually without rigorous surface preparation. However, this is not so for variants in lap and butt joints. Here performance depends on: basic metal or alloy; modulus; joint geometry; mode of loading and surface preparation. Steel alloys generally bond well and often, when toughened adhesives are used, do not require surface preparation. However, other common engineering metals and alloys can be difficult and surface preparation is often required to maximise performance. **15** Care – stressed plastic may crack. **16** Surface preparation may be needed. **17** Care – some forms particularly prone to stress cracking. **18** Requires special preparation for use in lap/butt joints. Rough or knurled surfaces usually acceptable for co-axial (slip-fitted) joints without chemical pre-treatment. **19** Some adhesives dissolve this plastic – take special care with the foam form. **20** Special problems – consult supplier. **21** Freshly cut, or lightly abraded, close fitting surfaces usually give the best results. **22** Butt joints are susceptible to damage and fail readily if abused. Extreme forms – such as bonded edges – must be avoided except in special circumstances eg honeycomb sandwich panels. **23** Care – absorption likely. **24** Apart from a few specialised adhesives (such as the polyimides – consult specialist supplier) – all conventional types decompose above this temperature. **25** Success depends on degree of modification – consult manufacturer. **26** A porous surface may be necessary. **27** Narrow bond lines may not cure properly. **28** Quick-cure versions are brittle. **29** High performance levels only from vulcanised systems. **30** Cyanoacrylates are brittle and readily degrade in warm wet conditions – particularly when metal surfaces involved. **31** Do not reject 'e' (cyanoacrylate) if: (a) it is a primary for both surfaces and, either (b) neither surface is metal or (c) one surface is a flexible rubber-based (30-43 inc) material. **32** Vapour extraction may be required.

How to use the Tables

1   Consult Table 5.3 (Adhesive/adherend compatibility) and, on a copy of Table 5.5 (Assessment), enter the status – R, S, P for reject, secondary and primary – of each adhesive for each adherend surface. 'Reject' here indicates a basic adherend/ adhesive incompatibility which renders the adhesive unsuitable. A primary adhesive/adherend combination will function well, right up to the limit of the adhesive's capability. 'Secondary' indicates that performance may not be so predictable. It should be appreciated, however, that an inherently high strength adhesive rated as secondary on a given adherend may well produce a far stronger bond than a weak adhesive rated as primary for the same material.

Note: if just one surface material is involved complete only one line. Also enter in the column provided the reference number(s) of the Special Notes given for each adherend material.

2   Examine the questions of Table 5.4 (Application questionnaire) and mark the relevant line and column with an x for each individual adhesive rejected. Note: often there will be a nil entry – no adhesive is rejected by the decision taken. Section 5.2.2 gives the background to each question to assist in interpretation.

Enter the reference number of any Special Note raised by a question and, in particular, take action on Note 31 in entering rejects.

3   Count the number of times each adhesive has been rejected – the number of x's on the completed questionnaire – and enter this number in the last row. Take particular care to log the zero entries – adhesives which have not been rejected at any stage of Table 5.4.

4   List all the adhesives which have suffered no more than two rejections in the consideration of Table 14. Note: exclude any adhesive rejected by Table 5.3 – ie an R in one of the adherend lines.

5   Consult Table 5.2 (Adhesive codes) and annotate each of the adhesives listed (4 above) with the relevant Special Note.

6   The adhesive(s) most likely to suit the application in question will have no rejections, but the final choice must take account of the significance of the Special Notes associated both with the adhesive(s) and the adherends involved (listed at the top of Table 5.5).

Sometimes, though not often, the Notes will indicate incompatability. Adhesives with no more than two rejections – provided they are relatively trivial ones – should then be considered. Alternatively, design modifications or production changes could be made to avoid the issue which caused the rejection.

If none of the less desirable adhesives produces an acceptable solution then a major design or production change becomes essential.

It can generally be said that the greater the number of adhesives allowed by the selection procedure, the greater the likelihood of finding a suitable individual adhesive for the application in question. The more restrictive the selection becomes, the less likely there is to be a satisfactory solution to the problem.

7   Finally, discuss the application with the adhesive manufacturer.

Table 5.4   Application questionnaire

Question Number	Questions	Yes/No	Reject	Go To
1	Do you intend to operate at temperatures above 220°C?	Yes	– See Special Note 24	
		No	–	2
2	Is the joint gap greater than 0.125 mm (0.005 in.)?	Yes	e	3
		No	–	4
3	Is the joint gap greater than 0.50 mm (0.020 in.)?	Yes	b c d e m n s t u v	4
		No	–	4
4	Is the joint co-axial ie composed of round, turned, threaded or fitted parts (Note: not axial butt joined parts)?	Yes	h o r	6
		No	c d	5
5	Is the width of the bond greater than 50 mm (2 in.)?	Yes	e	7
		No	–	7
6	Is the joint intended to be permanent – ie no possibility of dismantling for maintenance is acceptable?	Yes	r	8
		No	reject all but c d and k	Finish
7	Is the joint intended to be permanent – ie no possibility of dismantling for maintenance is acceptable?	Yes	r	8
		No	reject all but k p q r	8
8	If relevant will the joint be subject to peel, cleavage or impact forces? See Special Note No 22.	Yes	a b c d e g h l m n Special Note 31.	9
		No	–	10
9	Is the bonded assembly intended to flex readily and repeatedly in a manner that will distort the adhesive in the bond line?	Yes	reject all but f k p r s t w	10
		No	–	10
10	Do you require the adhesive to provide a positive gas or fluid seal?	Yes	a b e f l m r	11
		No	–	11
11	Will the joint be exposed to the weather?	Yes	a b e See Special Note 31.	12
		No	–	12
12	Are aggressive chemicals involved or will there be exposure to liquids above 90°C?	Yes	a b e f i j k p q r s t	13
		No	–	13
13	Is the cure time of the adhesive important?	Yes	–	14
		No	–	19
14	Is instantaneous adhesion on contact required?	Yes	reject all but f k q r	19
		No	–	15
15	Is it absolutely essential that handling strength be developed within 10 seconds of joint assembly?	Yes	reject all but b e k m n r	19
		No	–	16
16	Must handling strength be developed within 60 seconds of joint assembly?	Yes	a c f i j l o s u y z	19
		No	e k r	17
17	Must handling strength be developed between one and fifteen minutes after joint assembly?	Yes	e k l q r y	19
		No	–	18
18	Must handling strength be developed fifteen minutes or more after joint assembly?	Yes	b d e k q r t v	19
		No	–	19
19	Must the joint withstand a permanent shear load greater than 3.5 MN m^{-2} (500 psi)?	Yes	a b e f k q r s t	20
		No	–	20
20	Must occasional extremely high shear or rotational loads greater than 28 MN m^{-2} (4000 psi) be borne during use?	Yes	a b c d e f k l o p q r s t u v	21
		No	–	21
21	Are thin materials being bonded which may be distorted?	Yes	a b c d e g h l m n	22
		No	–	22
22	Is warming or heat curing possible even if not desirable?	Yes	–	23
		No	b g h j m n o x z	23
23	Are you prepared to use a two-part, mixed system?	Yes	–	24
		No	a b i j l m n p y z	24
24	Is the joint continuously exposed to water?	Yes	a b e f p s t See Special Note 31.	Finish
		No		Finish

Table 5.5   Assessment table

	a	b	c	d	e	f	g	h	i	j	k	l	m	n	o	p	q	r	s	t	u	v	w	x	y	z	Special note ref
Adherend 1																											
Adherend 2																											
Question 1																											
2																											
3																											
4																											
5																											
6																											
7																											
8																											
9																											
10																											
11																											
12																											
13																											
14																											
15																											
16																											
17																											
18																											
19																											
20																											
21																											
22																											
23																											
24																											
Total																											

5.2.2 The background to the selection questions

The questions of the selection procedure of Table 5.4 are as brief as possible, and while many of them are very precise, the following comments may help interpret others. They also provide background giving a useful insight into the whole selection process. The comments are presented in an order in which the appropriate questions are likely to arise.

Question 1
'Do you intend to operate at temperatures above 220°C?'
    Only very special adhesives cope with such temperatures and most will have lost much of their strength well below this temperature, beyond which oxidation and chemical degeneration become more severe. Seek a manufacturer of a polyimide adhesive as a first step if this threshold must be passed.

Question 4
'Is the joint co-axial, ie composed of round, turned, threaded or fitted parts?' (Note: not co-axial butt-jointed parts)
    Many adhesives are very difficult to use on co-axial assemblies.

Question 5
'Is the width of the bond greater than 50 mm?'
    Cyanoacrylates are difficult to handle when the bond width is greater than this, though joint length normally presents no problems.

Question 6/7
'Is the joint intended to be permanent with no possibility of dismantling for maintenance?'
    Of all adhesives, only anaerobics have been formulated to give reproducible and different levels of strength. They are the only adhesives which can be reliably used for the assembly of fitted, co-axial mechanical components where disassembly may be required for maintenance or other reasons. However, other types can be considered to be temporary and may be used in simple lap joints, e.g. adhesive tapes. A particular problem is presented here by adhesives which, in some applications, may reasonably be considered permanent – elsewhere their role is less readily defined. A particularly good example being the hot melt adhesives.

Question 8
'If relevant, will the joint be subject to peel, cleavage or impact forces?'
    Adhesives should ideally be loaded in shear and, if possible, combined with some compression. They should never be subject to peel, cleavage and impact forces but, in practice, some abuse may have to be tolerated, putting brittle adhesives at risk. If in doubt answer 'Yes'.

Question 9
'Is the bonded assembly intended to flex readily and repeatably in a manner that will distort the adhesive in the bond line?'
    The flexing of a bonded shoe sole illustrates the degree of distortion envisaged.

Question 10
'Do you require the adhesive to provide a positive gas or fluid seal?'
    For various reasons some adhesives do not form good seals. Only answer 'Yes' if the adhesive is really intended to function as a seal. Otherwise, the number of options may be unnecessarily restricted.

Question 11
'Will the joint be exposed to the weather?'
    Continuous outside exposure is meant here.

Question 12
'Are aggressive chemicals involved or will there be exposure to liquids above 90°C?'
    Cold acids and alkalis or any liquid above 90°C is meant here.

Question 13
'Is the cure time of the adhesive important?'

Think very carefully before proceeding either way – this point can very significantly affect the number of suitable adhesives.

Question 14
'Is instantaneous adhesion on contact required?'
    This is the sort of adhesion given by such materials as contact and impact adhesives, wallpaper paste and tapes etc.

Question 19
'Must the joint withstand a permanent load greater than 3.5 MPa?'
    Remember that this order of load may be generated when two stiff, dissimilar materials are exposed to elevated temperatures. Such an imposed load may be additional to the intended design load.

Question 20
'Must occasional extremely high shear or rotational loads greater than 28 MPa be borne during use?'
    Design loads this high should never be imposed for it is doubtful that any adhesive should ever bear loads greater than 20 per cent of its ultimate shear strength. Above this level, fatigue problems are likely, particularly in severe environmental conditions. However, such stresses may occasionally be reached during anticipated overload, or during accidental abuse. If in doubt answer 'Yes'.

Question 21
'Are thin materials being bonded which may be distorted?'
    Thin resilient materials – strip steel or semi-flexible plastics such as Perspex or unplasticised PVC – may be readily distorted without damage. However, even limited flexing may destroy brittle adhesives which would otherwise cope well. Again, if in doubt answer 'Yes'.

Question 22
'Is warming or heat curing possible even if not desirable?'
    Consider the possibility of gentle warming as well as heating to temperatures as high as 180°C – 200°C. Remember that sometimes heat-accelerated cure may be provided by another process or requirement, for example the heat of a paint stoving oven can be used to cure an adhesive.

Question 23
'Are you prepared to use a two-part, mixed system?'
    The emphasis here is on 'mixed' because this usually demands accurate manual work (and all that it implies) or expensive dispensing equipment. Remember that some two-part systems require no mixing, being activated by surface catalysis induced in turn by a primer pre-applied to at least one of the surfaces to be bonded.

6 Performance assessment

The performance of an adhesive needs to be assessed for three principal reasons:
>To see if it is suitable – both in production and in service;
>How it compares with other adhesives for the application;
>Quality assurance.

6.1 Basic suitability

The selection procedure of Section 5.2.1 should be used to identify families of adhesives most likely to suit an application. Manufacturers' data, which refer only to their specific products, should only be used to assist with the broader aspects of selection because no two commercial sources have a common interest in making available truly comparative data. And, even though the data may be strictly accurate, and based on standard tests, these frequently contain allowed variations affecting the results. Clearly, with differing organisations and unco-ordinated assessment, true comparison between different sets of commercial data is not possible. Another, and in many ways even more important, consideration besides how to compare data is what data are relevant. Until true design becomes technically possible, only an educated guess can be made regarding the desired performance characteristics.

This again makes the selection procedure of Section 5.2.1 appropriate for the first assessment.

6.2 Comparative performance

Historically, the comparative performance of adhesives has tended to be based on their shear strength. This is no longer good enough. Apart from production considerations, simple shear strength gives no indication of likely practical performance, which is almost certain to be determined by the ability to resist the combined onslaught of impact damage and environmental stress. Performance in these areas is probably best assessed by characterisation based on – impact, peel, combined impact/peel, wedge and shear stress tests all used in conjunction with various forms of environmental stress.

The following three useful, simple and practical rules will assist the final identification of individual adhesives:
>Choose an adhesive with high shear strength;

124

Check that peel and impact strengths of the adhesive are adequate;
Confirm that the environmental conditions will not cause an unacceptable
loss of performance in service.

But do not overlook the fact that standard tests use standard test pieces and
cannot compare as an assessment of suitability with the testing of the actual
bonded components themselves. A classic illustration is the torque testing (DTD
5628 Method D) of anaerobic adhesives compared to the actual vibration testing
of locked threaded fasteners (see Sections 2.3.2 and 3.6). The former test gives a
strength ranking – needed for design and quality control – while the latter shows
quite clearly that all these adhesives resist severe vibration – their raison d'etre.
The two are not comparable but are often confused.

If at all possible, take a long term view of adhesive performance. Use every
opportunity to monitor the performance of bonded components exposed to a
wide variety of field conditions. This ultimately helps correlate laboratory
assessment with actual performance. While manufacturers are always pleased to
offer evidence of performance, this is most unlikely to be strictly comparable.

So, the issue of testing is fundamental and must be left to those immediately
concerned, as only they know their aims and what the bonded components must
do. However, as inappropriate criteria may be selected because of lack of
experience, it is strongly recommended that discussions are held at the earliest
possible stage with a competent manufacturer to ensure best performance.

Characteristics such as adhesive moduli, other related properties and their
variation with temperature have not been mentioned – for two reasons. First, at
present, such data are sparse and, while crucial, no methodology is agreed for
their use or interpretation. Once again, the fact that design science has not yet
caught up with practical design experience has to be faced.

6.3 Quality assurance

Production considerations may dictate modification of certain of the selected
adhesive's properties to facilitate use. Once done, all the key characteristics
should be quantified and written up as a performance specification, which will also
serve as the purchasing specification, ensuring that materials are supplied to the
relevant standard.

Any good quality-assurance system should also assess the quality of
performance in practice as well as in theory. With no currently available, effective,
non-destructive test, critical areas or components may only be assessed for voids
(acoustic methods are useful here) and, where possible, proof loaded. Other than
this, test coupons or parts need to accompany the components themselves
through the assembly process so that appropriate, systematic destructive tests
may be carried out. A check-list for 'safety critical' items is given in Table 6.1.

The standard works describe individual test methods in great detail and the
more pertinent ASTM, BSI and MoD specifications are listed in Table 6.2.

Table 6.1   Quality assurance: checklist for critical structures

Adhesive production
- Type test to appropriate standard
- Batch acceptance using
     Shear, 'T' -peel, Impact, Viscosity and other appropriate parameters
- Periodic rotation through the type tests
- Constant QA assessment by suitably qualified external organisation (BSI – for example)
- QA assessment of raw material suppliers

Component production
- Assess newly delivered adhesives using batch tests
- Random stock checks (using batch acceptance tests)
- Periodic sequential rotation through type tests
- Constant monitoring of *all* aspects of surface preparation
- Wherever practical, employ application sensor on dispensing equipment
- Monitor bonded surfaces and ensure freedom from voids
- Prepare test coupons in parallel with actual components
- Constant monitoring of test coupon performance
- Carry out periodic destructive tests using actual components wherever possible

Table 6.2   Standard tests for performance assessment

Tests relating to properties of the adhesive
ASTM D1084-63 Viscosity of adhesives
ASTM D2183-69 Flow properties of adhesives
ASTM D1338-56 Working life of liquid or paste adhesives by consistency and bond strength

Tests emphasising tensile stresses
ASTM D897-72 Tensile properties of adhesive bonds
ASTM D2095-72 Tensile strength of adhesives by means of bar and rod specimens
ASTM D1344 Testing cross-lap specimens for tensile properties

Tests emphasising shear stresses
ASTM D1002-72 Strength properties of adhesives in shear by tension loading (metal to metal)
ASTM D3165-73 Strength properties of adhesives in shear by tension loading of laminated assemblies
ASTM D2182-72 Strength properties of metal-to-metal adhesives by compression loading (disk shear)
ASTM D2295-72 Strength properties of adhesive in shear by tension loading at elevated temperature (metal-to-metal)
ASTM D2557-72 Strength properties of adhesives in shear by tension loading in the temperature range $-267.8°$ to $-55°C$ ($-450°$ to $-67°F$)
ASTM D3164-73 Determining the strength of adhesively bonded plastic lap-shear sandwich joints in shear by tension loading
ASTM D3163-73 Determining the strength of adhesively bonded rigid plastic lap-shear joints in shear by tension loading
ASTM D3166-73 Fatigue properties of adhesives in shear by tension loading (metal-to-metal)
ASTM D2293-69 Creep properties of adhesives in shear by compression loading (metal-to-metal)
ASTM D1780-72 Conducting creep tests of metal-to-metal adhesives

Tests emphasising peel stresses
ASTM D903 Peel or stripping strength of adhesive bonds
ASTM D1876-72 Peel resistance of adhesives
ASTM D1781-70 Climbing drum peel test for adhesives
ASTM D3167-73T Floating roller peel resistance of adhesives

Tests employing cleavage stresses
ASTM D1062-72 Cleavage strength of metal-to-metal adhesive bonds
ASTM D950-72 Impact strength of adhesive bonds

Tests relating to bond durability
ASTM D1151-72 Effect of moisture and temperature on adhesive bonds
ASTM D1828-70 Atmospheric exposure of adhesive-bonded joints and structures
ASTM D3310-74 Determining corrosivity of adhesive materials
ASTM D1879-70 Exposure of adhesive specimens to high-energy radiation
ASTM D2918 Determining durability of adhesive joints stressed in peel
ASTM D2919-71 Determining durability of adhesive joints stressed in shear by tension loading
ASTM D3762 The Boeing wedge test

British Standards: A new British Standard dealing with a variety of adhesive test methods is currently in the course of publication. Interested readers should ask for the latest edition of BS 5350.
Anaerobic adhesives. Test methods for anaerobic adhesives may be obtained from the Ministry of Defence under DTD 5628 or from HMSO.
Cyanoacrylate adhesives. Test methods for cyanoacrylate adhesives may be obtained from the Ministry of Defence under TS 468 and TS 10168.

7 Manufacturing problems

Manufacturing problems occur with most industrial processes and adhesive assembly is no exception. However, adhesives are exceptional in that a lack of training and general unfamiliarity often make them a convenient scapegoat for any stoppage – while the real problem goes unnoticed and often unsought.

The same unfamiliarity tends to prevent original designs from incorporating adhesives; often they are pressed into service only when other production techniques are found wanting. This can occasionally put the adhesive critically close to its performance limit and minor variations in its use exagerate the effect on performance. This would not have occurred had the needs of the adhesive been taken into account in the initial design concept – when the requirements could have been properly accommodated. Thus, minor variations in production practice which can easily cause problems can be readily avoided if those concerned are aware of the likely pitfalls such as:

Is there enough adhesive?
   Can it be applied readily and accurately?
   Has it run away, dropped off or been forced out of position?
   Has the joint been made and re-made allowing the formation of voids?
Has the joint been moved prior to curing?
   Is it securely jigged and clamped?
Is the component as it should be?
   Have the dimensions changed significantly, even though they may still be within the mechanical design limits?
   Has the surface changed?
   are there differences in plating or passivation?
   has it oxidised, rusted or corroded in some way?
   has it become contaminated?
   Is the component still of the same basic material?
   Does the adhesive wet the surface?
   Is the component at its normal temperature?
Is the adhesive as it should be?
   Has its performance been independently checked?
   Has it been applied correctly?
   If mixing is necessary, has it been done correctly?
   Has it been contaminated?
   Is it the right adhesive?
   Has the viscosity/thixotropy altered?

If a primer is needed, has it been used, did it work?
Is the process as it should be?
    At the right temperature?
    Has the cure schedule been maintained?
    Are hot parts moved too soon?

As might be expected, the period during which an adhesive is introduced and a new production line established is the most critical and demanding phase of the operation. However, active liaison and co-operation between user and supplier should minimise and rapidly overcome teething problems.

Experience has shown that, with a fundamentally sound application and a correctly installed manufacturing process, a long period of trouble-free production can be anticipated after inauguration and once all concerned have become familiar with what is often, to them, a completely new production technology.

Appendix 1

Fundamental aspects of adhesion

K W Allen MSc CChem FRSC MInstP FPRI
Director of Adhesion Studies
The City University, London

Forces available

The whole aim of fastening two parts together, whether with adhesives, nuts and bolts, rivets or by welding, is to resist in-service forces trying to separate them. These forces are resisted by internally-generated forces in the joint and the characteristics of these must first be understood.

Ultimately the only available forces are the fundamental chemical bonding forces which hold all materials together. These are of several types:

Primary chemical bonds
a) Ionic bonds resulting from the Coulomb electrostatic attraction between particles of opposite charge.
b) Covalent bonds which arise from the sharing of paired electrons between essentially uncharged particles.
c) Metallic bonds, the consequence of the sharing in common of the free electrons in the conduction bonds of metal structures.

Secondary or van der Waals bonds
a) Dipole bonds (Keeson and Debye forces) arising from the uneven distribution of electrostatic charge in molecules and consequent polarities.
b) Dispersion forces (London forces) which are entirely universal between every pair of particles of molecular dimensions. They originate in the instantaneous asymmetry of charge distribution.

Hydrogen bonds
These are intermediate in properties between the primary and secondary bonds and depend upon the unique properties of hydrogen. They are of great importance in many situations including vital physiological processes.

The characteristics of these bonds are shown in the figure and table below.

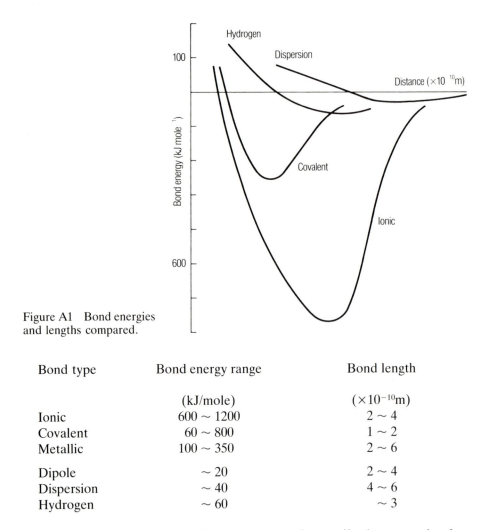

Figure A1   Bond energies
and lengths compared.

Bond type	Bond energy range	Bond length
	(kJ/mole)	($\times 10^{-10}$m)
Ionic	600 ~ 1200	2 ~ 4
Covalent	60 ~ 800	1 ~ 2
Metallic	100 ~ 350	2 ~ 6
Dipole	~ 20	2 ~ 4
Dispersion	~ 40	4 ~ 6
Hydrogen	~ 60	~ 3

The vital point is that these are all very short range forces effective over only a few (< 10) ångströms (1Å = 10^{-10}m), although the dispersion forces do extend rather further even if only weakly.

Surfaces

The roughness of real surfaces is very variable depending upon how they have been prepared. This roughness can be conveniently expressed as the average height of peaks above the median plane (ie half the average valley-to-peak distance).

Typical values for standard engineering finishes are:

Turned, milled, shaped $3 \sim 6\,\mu m$ ($1\mu m = 10^{-6}m$)
Drilled $2 \sim 5$
Ground $0.5 \sim 2.5$
Die cast $0.4 \sim 4.0$
Extruded $0.25 \sim 4.0$
Honed $0.1 \sim 0.5$
Lapped or polished $0.05 \sim 0.25$ ($500 \sim 2500$ Å)

Note particularly that even the very smallest of these values is a thousand fold greater than the ranges of action of the valence forces reviewed. So, on that scale, all are very rough indeed.

Putting two such rough surfaces together is like inverting the Alps onto the Himalayas. Certainly the surfaces cannot approach within bonding distance except perhaps just at the tips of the asperities. The only way a second material can approach sufficiently close to a sold surface is if it is a liquid, which can flow to establish intimate contact.

Exactly this technique is used with adhesives, which are always liquid at the application stage where temperature and pressure conditions promote flow and good contact.

Change of phase

The very property of flow, which enables a liquid to make intimate contact so that the available forces can have some effect, also prevents it sustaining any stress and it will always yield. So the property which makes forces available for adhesion also disqualifies it from acting as a useful adhesive.

But this apparent contradiction is surmounted by arranging that the applied liquid adhesive changes into a solid with properties appropriate to the demands of use. This phase change is crucial to the whole practical use of adhesives and may be brought about by various means:
(i) Loss of solvent (or dispersing agent), by evaporation and/or absorption in the substrate.
(ii) Solidification of molten material.
(iii) Chemical reaction of various types.

These three routes for the change of phase of an adhesive provide a very useful and common basis for classification of adhesives into: solvent- and water-based, hot melt and reactive types.

It may be noted that no consideration of pressure-sensitive adhesives is included in the present discussions. The relevant theory is quite separate and as the type of adhesive is of no significance in structural applications, it is omitted.

Wetting

Clearly the wetting of a solid surface by a liquid is crucial to satisfactory adhesion and merits careful attention.

Consider a motor car windscreen after a dry spell. Its surface is coated with a thin film of oil from exhaust gases mixed with traces of wax from cleaning and polishing. In the first shower of rain, the drops settle on it as separate and individual spots and, even if the rain is quite heavy, they remain in this form. Even with the windscreen wipers switched on the effect is smeared and blotched. Only the action of the washers with a small amount of added detergent spreads the rain into a thin and even film through which one can see satisfactorily.

This formation of a drop of liquid on a solid surface is described in scientific terms by the contact angle, θ, the angle between the solid surface and the tangent to the liquid surface at the point of contact.

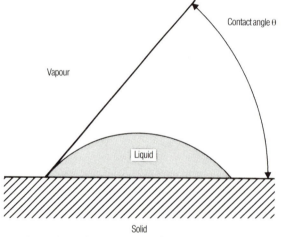

Figure A2   Contact angle between a liquid and a solid.

The value of θ can range from zero – complete wetting by the liquid – to 180° when the liquid remains as a spherical drop and does not wet the solid at all.

Although there is some discussion about the precise implications of the two extreme values of θ (0° and 180°) (complete spreading and total non-wetting), for all practical purposes these can be ignored.

Before proceeding with the main discussions about wetting, we should review the concepts of surface free-energy and surface tension.

Within the bulk of a liquid the attractive forces exerted on molecules by their neighbours balance in all directions; but at the surface this balance does not exist because there are no neighbours on one side (outside the surface) so molecules experience a net inward force. Thus to bring new molecules into the surface, work has to be done upon them, giving surface molecules a higher energy than those in the bulk.

It is this extra energy of the molecules in the surface which is called 'surface free-energy' or simply 'surface energy', expressed as energy per unit area, mJ m^{-2}. This is the energy needed to create a unit area of fresh surface.

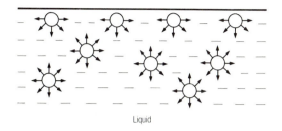

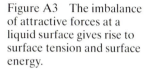

Figure A3   The imbalance of attractive forces at a liquid surface gives rise to surface tension and surface energy.

An older concept, arising from the same fundamental facts, is that of surface tension. The higher energy state of a liquid surface makes it behave as if it were under tension – as if constrained by an elastic membrane. This tension is expressed as the force exerted on a line in the surface in terms of force per unit length, mN m^{-1}.

Surface energy and surface tension are dimensionally equivalent and numerically the same, and the terms are commonly used almost interchangeably. Both are represented by gamma, γ. Since these are always associated with a surface, that is the boundary between two phases, it is sometimes necessary to specify which two phases – indicated by subscripts, eg γ_{12}.

Thomas Young first considered the wetting phenomenon in 1805 and he equated the surface tension vectors at the three-phase point of contact of solid, liquid and vapour and wrote:

$$\gamma_{SV} = \gamma_{SL} + \gamma_{LV} \cos \theta$$

which, while interesting and theoretically important, is of little practical value on its own.

The next step was taken by Dupré in 1869 who considered the work need to separate a layer of liquid from a solid surface

Work = energy of new surfaces created – energy of interface destroyed.

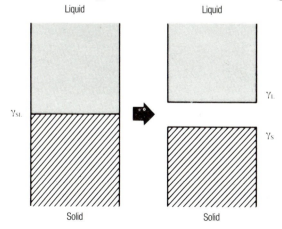

Figure A4   The physical representation of Dupré's analysis.

This is the Work of Adhesion, W_A

$$W_A = \gamma_S + \gamma_L - \gamma_{SL}$$

Notice the difference; in the Young equation the solid surface is in equilibrium with vapour and the corresponding energy is γ_{SV}; in the Dupré equation the separation gives a clean, bare solid surface and the energy is γ_S – better written, γ_{SO} to emphasise that the second phase is a vacuum.

The surface energy of a solid covered with a layer of adsorbed vapour is lower than that of a bare surface by an amount known as the 'spreading pressure', π, so

$$\gamma_{SV} = \gamma_{SO} - \pi$$

To make these two classic equations compatible, the Dupré equation is re-expressed, by substituting for γ_{SO}, as:

$$W_A = \gamma_{SV} + \pi + \gamma_{LV} - \gamma_{SL}$$

Now substituting for γ_{SV} from the Young equation gives:

$$W_A = \gamma_{LV}(1 + \cos\theta) + \pi$$

which accurately and properly describes the situation of a drop of liquid adhesive on a solid surface. Further, since surface energies do not change much on solidification, it reasonably represents a drop of solid adhesive on a surface.

π is always positive, although depending upon the natures of the vapour and solid may range from negligibly small (eg hydrocarbon oil on polyethylene) to considerable (eg water on metal oxides).

For a liquid, the work to divide it and produce two fresh surfaces, the Work of Cohesion, W_C, is clearly $2\gamma_{LV}$. Thus with perfect wetting ($\theta = 0$) then

$$W_A = 2\gamma_{LV} + \pi$$

$$= W_C + \pi$$

and so failures should always be cohesive within the adhesive rather than at the interface (i.e. adhesive failure).

Example
For a hydrocarbon liquid γ_{LV} is approximately $30\,mJ/m^{-2}$ and, if we assume that the inter-molecular forces become negligibly small over a distance 5×10^{-10}m, then the normal force required to pull the liquid from a solid and leave it covered with an adsorbed layer is

$$F = W_C/5 \times 10^{-10} = 2\gamma_{LV}/5 \times 10^{-10}$$

$$= 60/5 \times 10^{-10}\,mJ\,m^{-3}$$

$$= 1200\,kg\,cm^{-2}$$

This is much greater than the normal experimental values; for instance for polythylene/steel a failing load of $183\,kg\,cm^{-2}$ has been reported.

We shall examine the reasons for this large difference later.

The principle of minimum energy determines whether a liquid will spread on a solid surface; liquid will spread on a solid of greater surface energy but not on one of lower surface energy. Thus if a surface of high energy (the solid) is replaced by a surface of lower energy (the liquid) then the total energy of the system is reduced and this will be a spontaneous process.

i.e. $\gamma_L < \gamma_S$ liquid spreads

$\gamma_L > \gamma_S$ liquid will not spread

Various arguments to give this a quantitative basis have been advanced but most seem to contain inconsistencies and it rests most satisfactorily on this simple principle.

Solid surfaces are conveniently divided into two broad groups:

High-energy surfaces $\gamma_{SO} = 500 \sim 5000$ mJ m^{-2}, including metals and their oxides, silica, glass and diamond; generally hard materials of high melting point.

Low-energy surfaces $\gamma_{SO} = 5 \sim 100$ mJ m^{-2}, including most organic solids and polymeric materials; generally soft with low melting points.

Interestingly there appear to be no materials with intermediate surface energies.

Critical surface tension

Zisman studied the relationship between liquid surface energies and contact angles, using (initially) a homologous series of liquids and a low-energy surface. He found a linear relationship between the cosine of the contact angle and the surface energy of the liquid. If the line is extrapolated to $\cos\theta = 1$ then the surface energy corresponds to a liquid which will just, but only just, spread on the solid.

This surface energy is a characteristic of the solid surface and called the 'critical surface energy/tension', γ_C.

Zisman took care not to equate γ_C with either γ_{SO} or γ_{SV}, but regarded it as an experimentally measurable characteristic of the surface, related to the surface energy. Other workers have been less careful in this respect.

While for some surfaces it is ideal to use a homologous series of liquids, the actual possibilities are limited and a wider range of liquids have to be used. If these have polar character and especially if they are hydrogen-bonding, then deviations from simple linearity make extrapolation less satisfactory.

Leading on from the studies of contact angle and surface energy initiated by Zisman has been the work of Good and Girifalco.

This starts from the suggestion that the interfacial surface energy between a liquid and a solid should be given by:

$$\gamma_{LS} = \gamma_L + \gamma_S - 2\phi \, (\gamma_L \, \gamma_S)^{1/2}$$

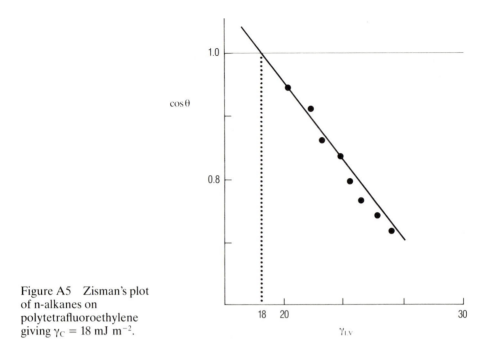

Figure A5   Zisman's plot
of n-alkanes on
polytetrafluoroethylene
giving $\gamma_C = 18$ mJ m^{-2}.

where ϕ is a function that can be calculated from the materials' thermodynamic properties. Hence it can be shown that

$$\gamma_C = \phi^2 \gamma_S$$

Fowkes went on to separate the forces acting across an interface and proposed that only those forces common to the two phases would be effective, this gives $\gamma_{AB} = \gamma_A + \gamma_B - 2(\gamma^d_A \gamma^d_B)^{1/2}$ where γ^d is the contribution of dispersion forces to surface energy. For a hydrocarbon, the surface energy arises solely from dispersion forces, while for mercury the surface energy involves both dispersion and polar, metallic bonding forces. So if A is hydrocarbon, $\gamma_A = \gamma^d_A$ but if B is mercury $\gamma_B \neq \gamma^d_B$ and $\gamma_{AB} = \gamma_A + \gamma_B - 2(\gamma_A \gamma_B{}^d)^{1/2}$

If this is combined with the Young equation for a non polar liquid on a solid, the dispersion contribution to the surface energy of the solid can be expressed.

$$(\gamma^d_{SO})^{1/2} = [\gamma_{LV} (1 + \cos\theta) + \pi] \div [2 (\gamma_{LV})^{1/2}]$$

and where π is negligible this dispersion contribution can be calculated since all the other terms can be measured.

Developments of this approach have proved to be particularly useful in exploring surface energy conditions.

Explanation of adhesion

So far we have examined the forces available for adhesion and the relationships between solid surfaces and liquids in contact. These must be brought together to explain the phenomenon of adhesion.

We will review each of the four general theories of adhesion which have been advanced: mechanical, adsorption, diffusion and electrostatic.

Mechanical

This is the oldest, simplest theory which in essence suggests interlocking. It underlies the layman's instinctive procedure of roughening surfaces to improve adhesion. Any proper consideration of quantitative data shows that this cannot be the true general explanation, although there are special cases where it is significant.

When adhesively bonding leather, it is important to roughen the surface to raise the fibres of the corium and for the adhesive to surround and embed them. Similarly, in the adhesion between textiles and rubber (of great importance in the construction of motor vehicle tyres) the extent of embedding of fibres of the staple yarn is the governing factor in achieving strong bonds. In both cases penetration of adhesive into the structure is no substitute for penetration of the fibres into the adhesive.

More recently with metal (strictly metal oxide) surfaces, evidence indicates that a very much smaller scale of roughness or porosity is important. Provided that the surface itself is strong and coherent, then a roughness with fibres or pores of micron diameter will increase strength in an adhesive bond.

Adsorption

This essentially depends upon the utilisation of the surface forces. Provided that the molecules of the adhesive and of the adherend can be brought close enough together, then the van der Waals forces will give rise to physical adsorption. This phenomenon of adsorption has been widely explored in pure physical chemistry and is quite well understood in simple cases.

It can be shown that the forces of physical adsorption are adequate for much more than the observed strength of adhesive bonds.

It is also certain that, in many cases, there is the possibility of chemical interaction and bonding across the interface. Electron donor-acceptor bonds may be formed to add to the adhesive strength from the dispersion forces. Hydrogen bonding is a particular case but acid-base interactions may also be involved.

Diffusion

This is largely due to a school of Russian chemistry lead by Voyutskii and Vasenin and is largely accepted as valid in adhesion between two surfaces of the same polymer and for the heat sealing of thermoplastic materials.

The concept is quite simple, one end of the polymer molecule chain from one surface diffuses into the structure of the second surface so that that molecule forms a bridge or bond across the interface. Using theories of diffusion and polymer structure, this has been developed to provide a substantial theoretical background which correlates quite well with experimental results.

This explanation is not only relevant in the adhesion of a material to itself, 'autohesion', but also for the technique of combining two films of adhesive above their glass transition temperatures. Many adhesives are commonly applied as films from solution or emulsion and are allowed to dry before the two surfaces are combined to form a bond – a diffusion process which removes any identifiable interface.

Electrostatic

This theory was developed by another Russian, Deryagin, particularly for pressure-sensitive tape. The adhesive and the adherend are likened to the two plates of a capacitor, and the work of separation is equated to that required to separate the two charged capacitor plates. Again the theoretical basis developed correlates with experiment.

This theory is not widely accepted as of general importance, but the mechanism probably contributes to certain rather special instances of adhesion.

Weakness

So far most of the discussion has centred on mechanisms of making adhesive bonds and their various details. It has already been indicated that real joints are nothing like as strong as theories suggest – indeed discrepancies are of several orders of magnitude.

The first cause of weakness is the existence of 'weak boundary layers', a concept first introduced by Bikerman who suggested that the idea that two phases in contact were entirely isotropic and of uniform composition up to their boundaries was mistaken. The surface regions of real materials are certainly different from their bulk and form the weak layers because of:
(i) The inclusion of air (or other gases) in voids or as bubbles within the bonding region.
(ii) The formation of oxidised layers – either themselves mechanically weak or only loosely attached to the underlying bulk material.
(iii) The concentration at the surface of minority constituents such as pasticisers, antioxidants, diluents etc, which are frequently either liquids or very soft solids.

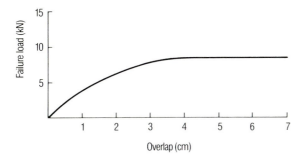

Figure A6   An increasing overlap length of a simple joint increases the failure load up to a critical maximum.

The second cause of weakness stems from stress concentration in a joint distributing it over a very much smaller area than might be expected from superficial consideration. This is illustrated by results for single lap joints where an increase in length beyond a fairly modest value has no effect on failure load.

A third cause of weakness is that, in all materials, surface irregularities provide the initiation points for crack growth. This follows the classifical theories of Griffith for metals and brittle materials, extended to tough materials by Irwin. These showed that there is a critical length of a crack beyond which it grows rapidly and leads to catastrophic failure.

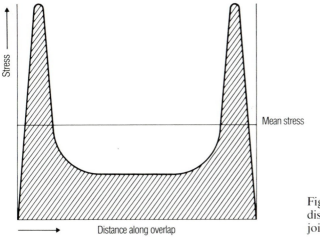
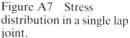

Figure A7   Stress distribution in a single lap joint.

A guide to the resistance of anaerobic adhesives to corrosive and aggressive media

The vast majority of industrial chemicals have little or no deleterious affects on hardened anaerobic adhesives and all grades can be used with them with little fear of compatability problems. However, the more aggressive materials such as the following, which have been selected from over a thousand industrial liquids and gases, need special consideration.

Up to a maximum concentration of 10 per cent, Group A materials below may be used with any grade of anaerobic to DTD's 5630-33 inclusive. Above this level the stronger grades of DTD's 5632 and 5633 must be evaluated. The final choice will depend largely upon component design and use.

Only DTD 5633 grades should be considered with Group B chemicals. With the exception of the materials given below, anaerobic adhesives should not be used with either Group C reagents or hot Group B chemicals.

Acetic acid 10 per cent	A		Chromic acid 50 per cent (cold)	B
Acetic acid – glacial	B		Chromic acid 50 per cent (hot)	C
Acetic anydride	B		Citric acid 5-10 per cent	
Acetyl chloride	A		(hot & cold)	B
Ammonia anhydrous	B		Formaldehyde (hot)	B
Ammonia solutions	B		Formic acid	B
Aqua regia	C		Gallic acid 5 per cent	A
Barium acetate and hydroxide	A		Hydrobromic acid	A
Battery acid	A		Hydrochloric acid	A
Bromine	C		Hydrofluoric acid	C
Bromine solution	B		Hydrogen fluoride	A
Bromo benzene	B		Hydrogen Peroxide (conc.)	B
Butyric acid	A		Lox (Liquid O_2)	C
Carbolic acid (Phenol)	A		Lye	C
Carbonic acid	A		Nitric acid 10 per cent	A
Chlorine dioxide	B		Nitric acid 20 per cent	B
Chlorine liquid	C		Nitric acid fuming	C
Chlorine (dry)	C		Oleum	C
Chlorosulphonic acid	C		Ozone	C
Chromic acid up to 30 per cent	A		Perchloric acid	A

Permanganic acid	C	up to 75 per cent (cold)	B
Persulphuric acid	A	up to 20 per cent (hot)	C
Phenol	A	Peroxides	C
Phosphoric acid		Sodium cyanide	A
up to 50 per cent hot and cold	B	Sulphonic acids	A
up to 80 per cent (cold)	B	Sulphuric acid (cold) to 70 per cent	B
up to 80 per cent (hot)	C	Sulphuric acid (hot)	C
Polyphosphoric acid	A	Sulphurous acid	B
Potassium cyanide	A	Tannic acid (cold)	B
Potassium and sodium hydroxide		Toluene sulphonic acid	B

Appendix 3
Unit conversion

1 lb f = 4.46 N

Impact
kpcm = kg f cm
1 kg f cm = 9.806×10^{-2} J
1 ft lb f = 1.356 J
1 kg f cm = 7.233×10^{-2} ft lb f

Shear
1 lb f in^{-2} = 0.0069 MNm^{-2} = 0.0703 kg f cm^{-2}
1 MNm^{-2} = 145 lb f in^{-2} = 10.197 kg f cm^{-2}
10.197 kg f cm^{-2} = 1 MPa
1 kg f cm^{-2} = 0.0981 MNm^{-2} = 14.223 lb f in^{-2}

Torque
1 lb f in = 0.113 Nm = 1.153 kg f cm
1 Nm = 8.85 lb f in = 10.197 kg f cm = 0.7375 ft lb f
1 kg f cm = 0.876 lb f in = 0.0981 Nm

Index